DARK
SKY
LEGACY

DARK SKY LEGACY

GEORGE REED

Astronomy's Impact on the History of Culture

PROMETHEUS BOOKS

Buffalo, New York

Library of Congress Cataloging-in-Publication Data

Reed, George, 1939-
    Dark sky legacy : the influence of astronomy on culture / George
Reed.
        p.   cm.
    Bibliography: P.
    ISBN 0-87975-541-5
    1. Astronomy. 2. Astronomy—Social aspects. I. Title.
QB43.2.R44   1989
520—dc20   89-10314
                                                              CIP

To my wife Joan,
the best thing in my universe

Contents

Preface

This book was begun twelve years ago when I began to write and illustrate a weekly "Naked i Astronomy" newspaper column for the *Daily Local News of Chester County*. The column is written for anyone who has ever looked at the night sky and wished they knew more about it. It developed into a blend of modern science, ancient science, mythology, history, linguistics, literature, and, of course, naked-eye astronomy.

In the process of writing the column, I began to discover that the affairs of our life were more greatly influenced by the night sky than I had imagined. It is impossible to adequately describe all that I learned in the process of creating a new column each week. Astronomy seemed to be connected to everything. The material I had accumulated began to cry out for some kind of organization.

I began to organize the material in 1985 when I was selected to participate as a faculty member in a Young Presidents' Organization university in Bangkok, where I was asked to speak on the astronomical and astrological traditions of the Eastern and Western hemispheres. It was in preparation for these lectures that I began to understand the strong connection between astrology and Eastern cultures.

The Bangkok lectures provided the material for an article, "Astronomy and Astrology—East and West," which appeared in *Astronomy* magazine in January 1986. The article proved to be not enough. I felt the need to expand and add to it. And so I wrote this book.

There are numerous people to acknowledge and thank for their contributions to the above events. These people include editors at the *Daily Local News,* especially Dennis Roussey, for allowing me to continue to learn about the influence of astronomy on culture through a weekly newspaper column; David Wechsler, the Education Co-Chairman of the 1985 YPO Bangkok university, for inviting me to become a participant in a first-class experience; and Richard Berry, editor-in-chief of *Astronomy* magazine, for accepting the "East-West" article and especially for adding color to my cartoons.

A very special thanks is extended to Paul Kurtz, editor-in-chief of Prometheus Books, for his decision to put the ideas expressed in this book into print.

And then there are the people who have listened to me talk about this material over the years. Their patience has often helped me focus my ideas. From West Chester University, these people include Dr. Sterling Murray, Lucy Troupe, and all the students who have passed through my astronomy and history of astronomy courses. I am especially indebted to the chairman of the Geology and Astronomy Department, Dr. Sandie Prichard, for the freedom she has allowed me to pursue my interests and take advantage of the opportunities that have come my way.

My long association with Spitz, Inc. has provided me with other people to thank for their support in this endeavor. These people include Charles Holmes, Mona Coldiron, Mike Bruno, Glenn Smith, and Hugh Sutherland.

Jack Spoehr of Imax, Inc. deserves special thanks for his continued enthusiastic support of every project I attempt.

And, finally, there is my family: my wife Joan and my children Christine, Tara, and Gary. To this group I now add Christine's new husband, Lieutenant Michael Cavanaugh, USMC. Thanks again gang. Your patience and humor has helped to even out the highs and lows of trying to use words to communicate my interest in our *Dark Sky Legacy.*

1
Lockyer's Key

"People only see what they are prepared to see." No one knows if the nineteenth-century American essayist and poet Ralph Waldo Emerson was thinking of the night sky when he wrote this but it certainly applies.

"Prepared to see" is the condition of being knowledgeable or educated in a particular subject. It is a form of literacy. Most often we think of literacy in terms of an ability to read and write. But there can be other forms of literacy. There can be as many forms of literacy as there are subjects about which to be informed. One of these, astronomical visual literacy, directly relates to the dark, star-filled, night sky. Astronomical visual literacy can be defined as an awareness of, and the ability to correctly interpret, naked-eye celestial phenomena.

As a culture, we have successfully sent men to the moon and brought them back, along with geologic samples of earth's closest celestial neighbor. But we are still collective failures at understanding the universe in which we live.

Part of the problem lies in the natural vanity of the human mind. We view the universe egocentrically, as if we personally were the center and essence of time and space. We tend to value most those things that can be seen to bear directly upon our lives.

It is the purpose of education to expand this narrow focus and provide a more cosmic view of the human experience. Astronomy and the night sky are subjects that satisfy this need. But they

are subjects in which our educational system has historically functioned well below the old "gentleman's grade of C" in terms of
creating a general astronomical visual literacy.

What should educated and visually literate people know about
astronomy and the night sky? At the very beginning of the twentieth
century, American astronomer Asaph Hall (1829–1907) expressed
his opinion. Hall was then the director of the United States Naval
Observatory in Washington, D.C. He was well-known as the astronomer who discovered Deimos and Phobos, the two natural
satellites, or moons, of Mars.

Hall included the following among his "common, everyday
facts of the science [astronomy] which can be learned by any
intelligent student without mathematical training": (1) why the stars
rise and set, (2) the motions of the planets and the moon among
the stars, (3) the reasons for our different seasons, (4) the names
of the principal constellations, and (5) why the constellations seem
to change with the seasons.

Hall reasoned that these aspects of astronomy and the night
sky are before our eyes all the time, and that everyone who is
fairly well educated ought to know something about them. He
also advocated the inclusion of aspects of the history of astronomy
in the curriculum of our nation's high schools and colleges.

Hall's lobbying efforts on behalf of astronomy education would
not have been needed in the nineteenth century. Astronomy was
then an integral part of the "natural philosophy" component of
the high school and college curriculum in the United States. But
between 1895 and 1910, there was a 90 percent decline in the
number of entering college freshmen who had studied astronomy.
In 1930, astronomy was being studied by only 0.06 percent of
all high school students. What had happened?

Astronomy had been dropped as a requirement for college
entrance and was soon eliminated from the high school curriculum.
It was also eliminated from the college curriculum. Biology, chemistry, and physics became the traditional "core" sciences.

Astronomy education experienced a renaissance in the decades following the launch of the Soviet Sputnik satellite in 1957, but this renaissance has fallen short of producing a population conversant with the "common everyday facts" that Hall, at the beginning of the twentieth century, thought educated people should know about astronomy and the night sky. The unfortunate truth is that our pre-space age ancestors of one hundred years ago had a greater awareness and understanding of the night sky than we do today.

The "common everyday facts" have been virtually eliminated from our educational system. And now they are being slowly eroded from our skies. We face the prospect of totally losing the visual stimuli, in populated urban areas, that have richly influenced our culture. We are in danger of losing a precious legacy from the past. The irony of the situation is that we are in danger of losing it by our own actions.

Pollution comes in many forms. Air pollution and water pollution have attracted the greatest attention because they are the most life-threatening forms of pollution. There is another type of pollution, however, that is also all around us. It is not life-threatening, but it is so bad that you can see it with your eyes. It's called light pollution.

Pollution can be defined as contamination by the discharge of undesirable substances into the environment. Light pollution arises from the discharge of excessive artificial light into the night sky.

Sky glow is a term used to describe light pollution. It is a measure of the brightness of the sky against the natural light of a dark, pristine night sky. The true night sky is assigned a value of 1. A sky glow of 1.5 means that artificial light has increased the sky brightness by 50 percent.

Some of the major astronomical observatories in the United States, once situated under pristine night skies, are experiencing serious problems because of the light pollution that results from increasingly populated and expanding urban areas. One major observatory has closed.

The Mount Wilson Observatory outside Los Angeles was once the envy of astronomers around the world. It possessed a 100-inch telescope, the largest in the world from 1919 until 1947. It was closed in 1986 because sky glow had made it practically useless for astronomical research. Light pollution had reduced the 100-inch telescope's efficiency to detect faint, distant objects to 11 percent of what it would be in a light- and sky-pollution-free environment.

The same fate may await the Mount Palomar Observatory, outside San Diego. Mount Palomar is the home of a 200-inch telescope, the largest in the United States. It is operating at an estimated 40 percent efficiency.

Our urban night sky is operating at about the same efficiency as the Mount Wilson Observatory. And the Mount Wilson Observatory was closed. Will our night sky someday be closed to us?

The human eye can supposedly detect four thousand to five thousand stars under ideal conditions. In one of today's "dark" residential areas, four to five hundred visible stars can create a religious experience. In a shopping mall parking lot, forty to fifty visible stars constitute a miracle.

Only the elimination of civilization would completely eliminate light pollution, and this would be a case of overreacting. But there are some things that can be done short of turning off all the lights.

Low-pressure, sodium vapor street lighting, for instance, is preferable to mercury vapor lighting because of its efficiency. The sodium vapor provides the maximum amount of its light in the narrow band of wavelengths to which the eye is most sensitive, and they contribute the least amount to sky glow. The simple shielding of light sources to direct light downward and prohibit light from flooding the skies would also be of benefit.

And then there is the question of unnecessary lighting altogether. Why, for instance, should an empty shopping mall parking lot be lighted after closing hours? Why should an empty sports stadium be lighted when not in use? Why should anything be lighted after its hours of use? Security is the answer most often given. But this answer has been criticized since high crime areas are usually also highly illuminated areas. It's another version of "Which came first, the chicken or the egg?"

The star-filled, night sky is a natural resource for the mind, and as such it is important that it be preserved. Planetariums and space theaters should not become astronomical zoos.

Do we need a dark, star-filled night sky? Yes! There is a part of our psyche that thrives on the mystery and majesty of the dark night sky. The dark night sky compels us to ask the harder questions about our lives. It compels us to ask questions about ourselves that no earth-based stimuli can provoke.

We look out at the wide expanse of universe beyond the earth, for instance, and ask a fundamentally important question about ourselves. It's a question that would never occur to us from our experiences on earth alone. We look into the night sky and ask, "Are we alone in the universe?"

Nobody answers.

From such a response, we may begin to conclude we are "special." We may begin to see our lives as the focal point within a grand design. But are we really special? Or are we allowing our egocentrism too full a reign? The constantly evolving answer to this question is woven into the fabric of the religious history of civilization.

The conditions for life on earth are perfect. And life has responded well. It would be difficult to ask the question "Are we alone?" on earth. The response would be deafening. Life on earth has evolved to thrive in an unimaginable multitude of forms and conditions. Some people may see this as the result of the grand design.

They argue that the sun is just the proper kind of star to allow sufficient time for life to evolve. And they are correct.

A star much hotter than the sun would produce too much of the high-energy radiation detrimental to life. A star much cooler would supply too little heat.

A star much more massive than the sun would evolve too quickly. The supernova death of the star would destroy the conditions for life on a planet long before life had sufficient time to evolve.

A star much older than the sun would have formed too early in the history of the universe. It would not have been able to supply a planet with the necessary life-based elements, such as carbon and oxygen. These elements are produced in the supernova explosions of other stars. It takes time for the supernova-produced elements to be distributed to the nebular clouds of later forming stars and planetary systems.

The emergence of life on a planet in orbit about a star is not an accidental event. It requires timing, conditions, and a stability measured in billions of years. It requires, first of all, a star like our sun.

The grand design proponents also argue that the earth is just the right kind of planet to allow life to evolve. And they are correct again.

The earth is situated in the ecosphere, or life zone, of the sun. This is a region surrounding the sun where temperatures permit water to exist in liquid form. Water is the bloodstream of evolution. It is the mechanism by which the right things are in the right place at the right time for life to happen.

A planet much closer to the sun, like Mercury with daytime temperatures of 800 degrees Fahrenheit, would be too hot for water to exist as a liquid. A planet much farther from the sun, like Mars with nighttime temperatures of minus 126 degrees Fahrenheit, would be too cold for water to exist as a liquid.

A planet in the ecosphere much more massive than the earth would not be supportive of earth-type life. Its gravity would most likely maintain a thick, poisonous atmosphere and possibly create the runaway "greenhouse" effect of Venus. A planet, like our moon, with too little mass would be unable to hold an atmosphere. Liquid water would evaporate into space.

A planet formed within a multiple star system, instead of a single star system like our solar system, would have little chance of supporting the evolution of life. The planet's orbit would most probably be unstable, and the planet would eventually be ingested or ejected by one of its suns.

And, the grand design advocates point out, even the simplest life-form is not that simple. A typical bacterium cell may be composed of as many as one hundred million molecules.

Life involves the extremely complicated ability to take from the environment what is needed for growth, survival, and reproduction. Can this just happen accidentally? Or does it need a little "push"?

We look out and ask, "Are we alone in the universe?" And nobody answers. But does this truly mean we are "special"? Does this truly mean we are the product of a grand design? Or could it mean we are able to ask the question simply because statistical probability produced a little planet in the proper orbit, around the proper star, at the proper time?

The answer to the question of our "aloneness" is fundamentally important in determining if we are "special." The answer to the question of our "specialness" is fundamentally important to all other questions.

If we ever receive a reply, I hope it will be something other than: "I'm sorry but we are not able to come to the phone right now. But if you will leave your name and telephone number at the sound of the tone, we will get back to you."

There is another hard question that the dark night sky compels us to ask. It is the question of how the universe was formed. Where did it come from? Is the universe "special"? This is a dark night sky question that has not been confined to the dialogue among astronomers alone. Even Huck and Jim, in Mark Twain's *The Adventures of Huckleberry Finn,* have joined the debate:

It's lovely to live on a raft. We had the sky up there, all speckled with stars, and we used to lay on our backs and look up at them, and discuss about whether they was made or only just happened. Jim he allowed they was made, but I allowed they happened; I judged it would have took too long to make so many.

Cosmology is the study of the nature, origin, and evolution of the universe. The twentieth century's answer to cosmology's fundamental question of whether the universe "was made" or "only just happened" is "yes." It was made from an event that only just happened—the Big Bang creation of the universe some fifteen to twenty billion years ago.

The expanding matter of the cosmic fireball that created space and time cooled to form the clusters of galaxies that we observe to be still expanding today. The earth, our moon, the planets of the solar system, and we ourselves, are all by-products of the formation of the sun within the Milky Way galaxy. The solar system was formed from the accretion of material in the gas and dust surrounding the early sun. Since we are participants within this expanding universe, we observe its results in the Doppler effect, redshifts of the galaxies in all directions of space.

But still we ask, did this just happen accidentally? Or did it need a little "push"? Again, we ask if we are the product of a grand design or the result of a statistical probability. Again, we realize this is the most important question we can ask. Its answer affects the answers to all other questions.

Would people still ask the question if there were no dark night sky? I doubt it.

Our earth-based dark sky visual literacy is also in danger from another source: the future. It is in danger of being forgotten in future events involving the exploration and development of the space frontier. Part of the "old country" will inevitably be left behind as science, technology, and enterprise make accessible new resources to support human settlements in and beyond earth's orbit. Progress and change have their casualties.

Our future in space is inherited from the dark night sky. It is part of the human desire to travel beyond the horizons of our vision. Where once we left the earth in the creations of fiction writers, we will soon leave the earth in the creations of science and technology.

It is difficult to imagine the civilian space goals of the 1986 Report of the National Commission on Space as a reality of just fifty years into the future. The world they foresee a half century into the future is a step into science fiction in comparison to the world of today, and a leap into a crazy man's dream in comparison to the world of half a century ago.

Imagine the Mars that appears as a reddish star in our night sky as the home and workplace of people living in self-sustaining biosphere bases on the Martian surface or on Phobos, one of the red planet's two natural satellites.

The biosphere base you are imagining will be an enclosed ecological system that supports and maintains itself. The survival and health of the human "Martians" will necessarily require air, water, food, and protection from cosmic and solar radiation. This will require entirely new technologies and research into the long-term effects of low gravity on human beings.

The routine trip to Mars will take place aboard a routine orbiting spaceship between earth and Mars. The half-year voyages out will bring new crews and cargo to Mars in exchange for old crews and cargo to complete the cycle on the half-year return voyages to earth.

What will they be doing on Mars? The time-honored answers are predictably repeated by the Commission. They will be searching for evidence of extraterrestrial life ("Are we special?"), and they will be attempting to understand the creation and evolution of the universe. ("Is there a grand design or did it just happen?" Maybe. But this will probably be the secondary reason for settlements on Mars.)

It was President Calvin Coolidge who said, "The business of America is business." The Commission agreed. The United States is, and will continue to be, in the space business.

The Commission sees the inner solar system as a space frontier rich in raw materials. They recommend the use of remote-sensing, robotic explorers and prospectors, and, finally, where necessary, human beings to seek and retrieve valuable mineral resources from earth-crossing asteroids as well as the moon and Mars.

The mineral resources, the Commission predicts, will go into the manufacture of spaceports, space stations, moon bases, Mars bases, cycling spaceships, cargo transport vehicles, passenger transport vehicles, communication satellites, and navigation satellites.

Space enterprises using robotic factories can be expected to provide solar energy for earth consumption, propulsion fuels, and exotic new products that can only be produced in microgravity, a vacuum, or low-temperature environments. It is conceivable that future generations will be plagued with a deficit balance of payments due to the importation of "Made in Space" products. The call of the future may be "Save Jobs, Buy From Earth."

It is hard to imagine the sci-fi goals of the National Commission on Space as the reality of just half a century into the future. But think back to the world of just half a century ago. Pan American Airways was just opening the transpacific highway and was preparing to open the transatlantic highway with "flying boats."

If the vision of the National Commission on Space comes true, it will not be the first time a sci-fi vision of the future has evolved into reality.

On October 5, 1863, president Impey Barbicane announced to the members of the Baltimore Gun Club that he proposed to send a missile to the moon. A sum of 5.5 million dollars was subsequently raised for the project, and three adventurers blasted off in the following year on what amounted to a twelve-day exploratory voyage around the moon. The lunar explorers were treated as celebrated heroes upon their return to earth.

If you did not recognize the above as the fictional plot of Jules Verne's novel *From the Earth to the Moon,* you probably at least felt familiar with the series of events. It has often been stated that today's science fiction is tomorrow's science, but never has this been as true as in the case of the nineteenth-century French author Jules Verne.

Jules Verne's novel of a lunar voyage was science fiction that came so close to the actual events you could almost believe he had a clear vision of the future, or that NASA scientists had used his novel as the basic script for their own moon adventure.

Verne predicted that the first moon voyage would be an American project. He was right. Verne sent three adventurers on a four-day December flight to the moon. The Christmas 1968 voyage of Apollo 8, which was the first to place astronauts (William Anders, Frank Borman, and James A. Lowell) in lunar orbit, took three days from launch until arrival at their orbit around the moon.

Verne launched his astronauts from a 900-foot cannon, the Columbiad, buried in the earth near Tampa on the west coast of Florida. The Kennedy Space Center, from which Apollo 8 was launched, is located less than 120 miles away on the east coast of Florida.

Verne's astronauts were picked up by a United States Navy vessel that accidentally witnessed their spashdown in the Pacific Ocean southeast of Hawaii. The Apollo 8 landing was witnessed by the United States Navy's aircraft carrier *Yorktown.* It was four miles away and waiting to pick up the Apollo 8 crew. The similarities between fact and fiction go on.

But Jules Verne did not guess everything correctly. Verne did not include a system of communication between his astronauts and the earth. An attempt was made, however, to visually follow the voyage by the use of a 192-inch-diameter reflecting telescope that was especially built for that purpose in California. (The largest reflecting optical telescope in the United States, the two-hundred-inch Hale telescope, is located at Mount Palomar in California. It was completed in 1948.) The Apollo 8 spacecraft was constantly in radio communication with its ground control, by way of a radio telescope tracking system, except for the time when the Apollo 8 spacecraft was behind the moon. Verne's voyagers also lived somewhat better during their trip than the Apollo astronauts, who were not provided with a supply of fine wines and cigars.

With the expansion of the space frontier will come a continued acceleration in the knowledge explosion. As the space age evolved during the third quarter of the twentieth century, knowledge began to double at estimated seven-year intervals, with predictions of even shorter intervals to come.

It would be tragic if our awareness and understanding of the dark night sky were lost among the exponentially increasing nuts-and-bolts knowledge of science and technology. We might go to Mars, but we might also leave behind some of our humanity. We might forget the route by which we reached Mars.

Astronomy is the study of the universe. It involves questions of the where, what, and how of space and time. But it is impossible in considering these questions not to consider also the related fundamental question of "why?"

Astronomy evolved from the dark night sky. The dark night sky was the first universe. But astronomy is not an isolated study. Over the years, it has taken from and given to a wide range of disciplines. It has become a part of world culture. Astronomy has become connected to almost everything in some way. Sometimes the connection is obvious, sometimes it is not. And this is our *Dark Sky Legacy*.

One of the early investigators of the influence of the sky upon different cultures of the world was the English Victorian scientist

J. Norman Lockyer (1836–1920). It was Lockyer who provided the conceptual scheme that I have followed in this book.

The career of J. Norman Lockyer had two distinguishing characteristics, one of which was derived from a personality trait. Lockyer was a maverick. He had a tendency to be involved in controversy, and, in fact, as founder and editor of the weekly science journal, *Nature* magazine, he sought controversy. Controversy increased circulation. His other distinguishing characteristic was that he had a tendency to support positions that, while basically sound, were not totally accepted. Lockyer often had the correct key, but the wrong door.

Although born in Rugby, England, Lockyer was not the product of an English university but was educated on the Continent. In science, Lockyer was a self-educated man. In 1861, he became fascinated with a neighbor's refracting telescope and ordered one to be built for his own use. This was the beginning of his lifelong interest in astronomy, an interest which would consume his leisure time and personal resources. It was an interest that would earn him knighthood and honorary doctorates from Cambridge and Oxford universities. It was an interest that would lead to his being considered for the prestigious Savillian Chair in astronomy at Oxford.

Using a six-inch telescope and a small spectroscope, Lockyer began a study of the light coming from the dark sunspots found on the solar disk. As a result of his observations, in 1868 he announced that the violent solar prominences associated with sunspots were upheavals in a gaseous layer of the sun's atmosphere. This was a layer above the sun's bright, visible surface, the photosphere, and the large, tenuous corona surrounding the solar disk, visible only during solar eclipses. Lockyer gave this layer the name by which it is known today, the chromosphere.

As fate would have it, a French astronomer, P. J. Janssen, (1824–1907), working independently in India, announced the same results by letter at the same meeting of the Academy of Sciences in Paris. Both astronomers had discovered the same method of observing solar prominences without the need of traveling to remote places to view a minutes' long solar eclipse. Lockyer was the first to think of the method, but Janssen was the first to realize the application, so both names are connected with the discovery.

During a total solar eclipse visible in India that same year, Lockyer found a prominent yellow line in the spectrum of the solar atmosphere that could not be identified with any known element on earth. Lockyer assumed he had discovered an element in the sun that did not exist on earth. He named it "helium," the sun element, in honor of Helios, the Greek god of the sun.

Helium was discovered on earth twenty-seven years later by British chemist William Ramsay (1852–1916). Helium is the second-most abundant element in the universe. It is only surpassed in abundance by hydrogen. But it is a colorless, odorless, tasteless, inert gaseous element that is found in very few compounds, which explains the difficulty experienced in its discovery on earth.

In the year following the discovery of helium, 1869, Lockyer inaugurated *Nature,* a weekly science magazine. He remained its editor until his death more than fifty years later. The success of the journal was built on controversy. Through the inclusion of letters to the editor and anonymous book reviews, Lockyer fanned the fires of controversy surrounding the burning scientific issues of his day. Lockyer often wrote the controversial letters himself under different names. People read *Nature* whether they loved it or hated it.

Lockyer did not allow the editorship of *Nature* to interfere with his scientific pursuits. He remained an active researcher. He found time to attempt to solve contemporary difficulties in the interpretation of stellar spectra with his "dissociation hypothesis." Early in his studies of the chemistry of the sun, Lockyer postulated that the problems of spectroscopy could be resolved by "celestial dissociations." Lockyer wanted to expand the series of chemical dissociations from molecules to atoms to include a larger number of terms. He suggested that the chemical atom itself could be decomposed almost without limit if the temperature was sufficiently increased.

The nonacceptance of the dissociation of the atom by chemists of the time greatly frustrated Lockyer, and he continuously referred to the mistake they were making in applying the principles of low-temperature chemistry to high-temperature conditions.

Lockyer, unlike many pioneers, was fortunate to have lived to see his dissociation hypothesis gain scientific respectability. It came with the discovery of the electron and the acceptance of the Bohr model of the atom. In 1913, Danish physicist Niels Bohr (1885–1962) proposed that an atom was a central nucleus surrounded by electrons in restricted energy orbits. The Bohr model of the atom served as an explanation of the manner in which atoms absorb and emit energy. Lockyer's dissociation theory predicted the behavior of the Bohr atom under high temperatures. It could have easily been translated into the modern theory of ionization.

After 1890, Lockyer became interested in the chronology of history and the possible use of astronomy as a means of establishing historical time frames. He was drawn to Egypt, with its monuments and temples of undetermined age, to test his hypothesis. In 1894 he wrote and published *The Dawn of Astronomy,* a book based upon his investigations.

Like the dissociation hypothesis, *The Dawn of Astronomy* was greeted with skepticism and ridicule. Much of what Lockyer proposed has proved with time to be incorrect. But one of his proposals has provided an excellent conceptual framework within which to look at the cultural influence of astronomy.

In *The Dawn of Astronomy,* Lockyer proposed that the use of celestial objects and phenomena in different cultures can be divided into three distinct stages. He never insisted one stage followed another, but he did propose that at different times in the development of a culture, one stage could be more dominant than the others. This is what we observe when we trace the history of the influence of astronomy in different cultures. This is also what we observe when we compare the history of the influence of astronomy in the composite cultures of the Eastern and Western hemispheres.

Lockyer described his first stage as one of primitive wonder about the sky, and the worship of the objects of the sky. This is the sky of myth and magic, the sky of the ancient stargazers.

Evidence of this first stage can be found in the early history of almost every culture. The sky and its objects were tied to gods and myths, which served as explanations for the existence, structure, and creation of the universe. The sky myths often became a hidden repository for human knowledge.

Vestiges of this myth and magic sky still exist today in subtle and sometimes unrecognized form in the symbols and language of everyday life. Sky phenomena still influence our lives when they serve as the inspiration for song lyrics as well as artistic renderings. Sky phenomena are represented on the flags and stamps of the world. The sky of myth and magic has also provided the marketing world with impact words and has enriched our language with numerous terms and expressions. The history of Western civilization can be found in the names associated with the stars and objects of the heavens. Its representation is everywhere.

Lockyer's second stage in the development of astronomy involved the utilization of celestial phenomena to serve the practical needs of culture. No more practical need was served by astronomy than that of the establishment of time and a calendar. The sky alone offered the necesssary repeating cycles to establish measures of time. The sky is a clock and a calendar. Without the sky, without time, there would be no civilization.

Another practical use of the sky is for building in orientation with the horizon. The recent popularity of the challenging interdisciplinary field of archaeoastronomy has revealed numerous incidences of the use of astronomy in structures built by ancient cultures throughout the world. The alignment of structures with particular directions along the horizon served the purposes of calendrical observation, power display, and ritual. A strong case can be made for Lockyer as the founder of archaeoastronomy.

Astrology was also developed for totally practical purposes. It was intended as a predictive tool to be used in seeing the future. It was also an attempt to understand a confusing world of seemingly random events.

Astrology is an ancient tradition that arose from an attempt to utilize celestial objects and phenomena of the macrocosm as predictors of events in the microcosm. It is a pseudoscience that greatly differs from astronomy in method and intent, but is often confused with the science of astronomy. Astrology in its many forms is the subjective study and interpretation of a proposed relationship or interaction between man, events, and the universe, with special emphasis upon the cosmic events of the region of the sky called the zodiac.

Astrology, because it only deals with a possible parallelism between the timing of events in the cosmos and the individual consciousness, does not deal with causes and effects, as does the scientific pursuit of astronomy, but rather with indications of the possibilities or probabilities of certain events occurring in certain places at certain times. *"Astra inclinant, non necessitant"* is the ancient motto of the astrologers—"The planets incline, they do not determine." While astrology has been under attack for centuries in the West, a covert belief in celestial influences in human affairs still remains.

The Eastern and Western hemispheres differ drastically in the influence that astrology has had on the development of astronomy's third stage. Lockyer's third stage involves the observation and study of celestial objects for the acquisition and advancement of knowledge for its own sake, without consideration for its practical use.

Lockyer's third stage blossomed with the invention of the telescope in 1610. Astronomy became research-oriented and was sup-

ported by governments, educational institutions, and private
sources; it no longer had to sell its services through astrology to
survive. Astronomy continued to have some practical uses, but
these were in the realm of technological applications of advancing
knowledge.

A new understanding of the cosmos and the earth's position
in space and time was derived from the labors of Galileo, Kepler,
Newton, Halley, Herschel, Einstein, Hubble, and Penzias and
Wilson.

Missing from this list are names of important contributers
from the Eastern Hemisphere. The East was satisfied to stay locked
in the pursuit of astrology, while the West pursued the universe
as a subject of research. It has been only in the latter part of
the twentieth century that astronomical research has received support in the Eastern Hemisphere.

The difference between the development of Lockyer's three
stages in the East and the West may be found in the differences
in the religious and philosophical thinking of the two hemispheres.

Eastern religious and philosophical schools can be characterized as animistic in their world view. They ask "why" questions
and seek a designing intelligence beyond the supposed facade of
sense perception. They look for explanations of the universe in
terms of underlying intentions and purposes. Eastern religious and
philosophical thought seeks spiritual escape from a confining
dependence on the body and the physical world, and therefore
has little interest in encouraging comprehension of what it is trying to escape.

Western religious and philosophical schools, on the other hand,
can be characterized as mechanistic in their world view. They ask
"how" questions. They approach the universe in terms of physical
cause and effect. The Western astronomer studies the universe in
much the same way as a Western religious believer studies sacred
literature to know religious truth.

Lockyer's three stages give us the key to organize and examine the history and influence of yesterday's astronomy. They give us a way to put today into perspective.

The past is gone, never to return, but what the past has left behind belongs to the present and the future. It is our *Dark Sky Legacy*.

2

"Saving the Phenomena"

In the second century B.C., the Greek philosopher Plato gave his students the most challenging of assignments. He required them to "save the phenomena." He wanted them to devise a conceptual system that would reproduce the appearance and motions of the naked-eye day and night skies at any time in past, present, or future. The history of astronomy and horology (the science of measuring time) has been the history of continuing attempts to meet Plato's challenge.

The phenomena that was to be "saved" by Plato's students is what is called naked-eye astronomy, or the astronomy of the stargazers. It is dark sky astronomy. It is the astronomy of time-keeping and navigation, the astronomy of archaeoastronomy, the astronomy of astrology, and the mask that covers the universe beyond. This is the astronomy that has most infiltrated and influenced the cultures of the world.

When it comes to astronomy, it is surprising what we consider to be "natural" and "unnatural." This is especially true in terms of our view of space. The natural view of the solar system for many people is that of the sun surrounded by nine planets. This is the view one would have from above the plane of the solar system. No one, nor any unmanned spacecraft, has ever viewed the solar system from this vantage point in space. This view is not our natural view.

The word "natural" is often mistakenly used as a substitute

or synonym for the word "familiar." The view of the sun and planets as seen from above the plane of the solar system is the familiar view of the solar system, because the familiar view for many people is, unfortunately, the interpretative view that is presented in textbooks. The "real" natural view is what is learned from observing the "real" world, the "real" day and night sky. This is the phenomena Plato wanted saved.

If you take the opportunity to view a clear evening sky in reasonably dark surroundings, you will no doubt notice that the stars are of varying brightness. You may logically assume that the brighter stars are relatively close to you and the fainter stars are farther away. A closer inspection will reveal that it is impossible to verify this assumption. It is impossible to visually discern any differences in the distances of the stars; in fact, all stars appear to be at the same distance from the earth. All observers of the night sky, everywhere on earth, appear to be in the center of a large heavenly or celestial sphere that is bisected by the horizon.

The ancient stargazers were followers of the "seeing is believing" axiom. They believed in the reality of a celestial sphere. It was only with great difficulty that the concept was eventually abandoned.

We know today that the stars are really at varying distances from the earth and that the celestial sphere is only a fictitious sphere that appears to have stars fixed upon it. The concept, however, is still useful in understanding the motions of the stars and other celestial objects viewed by a terrestrial observer. The celestial sphere provides a conceptual economy for saving the phenomena.

The myriad stars seen in the evening sky present an endless possibility of convenient groupings. Somewhere in history, the brighter stars came to be named. Certain large groups of stars are called "constellations," which is derived from the Latin "with stars." Constellations provided a map to navigate the celestial sphere, a map that was culture-dependent.

The grouping of stars into constellations is a purely arbitrary process. Some of the constellation figures are without any recognizable pattern or association relating to their name. A few of the more commonly known star patterns, such as the Big Dipper and Little Dipper, are not constellations. They are asterisms. An asterism is a part of a larger constellation that contains a well-known and recognizable group of stars. The Big Dipper stars are part of the much larger constellation Ursa Major, the Large Bear. In the British Isles, the Big Dipper is known as the Plough. The Little Dipper stars are part of the larger constellation Ursa Minor, the Small Bear.

The origins of most constellations have been lost in antiquity, but each constellation figure has been given a name that is associated with some ancient myth, animal, or common object. Some names are very old because the Greeks and Western civilization adopted many of the earlier Babylonian constellation names and forms. Some constellation names are comparatively recent because they were created during and after the voyages of discovery by Europeans to the Southern Hemisphere. Some constellations were created simply to fill in the empty spaces containing only faint stars between the ancient constellations.

The mapping of the night sky by tradition created some confusion. Some stars belonged to different constellations on different maps. Some faint stars were not associated with any constellation. In 1930, the International Astronomical Union (IAU) solved the problem. They created a universally agreed-upon map of the celestial sphere by dividing it into eighty-eight official constellations. The IAU adopted boundary lines for the constellations that were outside and independent of the constellation figures.

The appearance of the dark night sky changes with the seasons and the years. It changes because of the earth's motions of rotation and revolution. And it changes because of the real motions of the moon and planets. These changes must be understood in order to understand the influence they have had upon world culture.

The stars and constellations of the celestial sphere appear to move from east to west with respect to the horizon as the night progresses. This causes the stars and constellations to appear to rise in the east and set in the west. This apparent motion of the celestial sphere is referred to as its diurnal or daily motion. That this was a true representation of reality was a serious belief from antiquity into the seventeenth century. The longevity of this concept is easy to understand. Your senses and intuition will tell you that you are stationary and it is the sky that is moving. The opposite is true. The motion is only an *apparent* motion due to the earth's rotational motion in the opposite direction.

The earth rotates counterclockwise 360 degrees, with respect to a given star on the celestial sphere, in 23 hours and 56 minutes. This is the measure of a sidereal (star) day. Since the sidereal day is four minutes shorter than the mean solar day (civil time), the stars will rise and set four minutes earlier each day. It also means the stars will drift slowly westward from night to night at the same time of observation. The constellations will make seasonal appearances.

The earth's axis of rotation passes through the terrestrial north and south poles. This axis can be extended to intersect the northern half of the celestial sphere at the north celestial pole, a position ¾ degrees away from the star Polaris. Polaris is the "pole star." The celestial sphere will appear to rotate around Polaris, and Polaris will appear to remain stationary. Since Polaris is always found in the true north direction, it has become known as the "North Star." Of great benefit to presatellite age travelers and explorers was the knowledge that the altitude of Polaris above the true north direction on the horizon is equal to the geocentric latitude of the observer.

The altitude of Polaris determines the number of circumpolar stars and constellations visible from any latitude. Circumpolar stars and constellations are those stars and constellations that are able to "circle the pole" without going below the horizon. Noncircumpolar stars and constellations are the seasonal stars and constellations. These are the stars and constellations that can be seen to rise and set on the horizon.

The rotation of the earth is also responsible for the apparent east to west daily motion of the sun with respect to the horizon. But the earth also revolves counterclockwise around the sun in a period of 365.25 days. This motion of the earth causes the sun to appear to move approximately one degree per day, counterclockwise, from west to east with respect to the stars on the celestial sphere.

The path of the sun through the fixed stars is called the ecliptic. The word comes from the Greek *ekleipsis*, which means "being absent." The ecliptic path is where eclipses or absences of the sun and moon take place. A series of twelve zodiac constellations lie along the ecliptic. Actually, there are thirteen constellations that are intersected by the ecliptic, but Ophiuchus the Snake Bearer is traditionally ignored because it is a relative latecomer to the twelve from ancient times. Zodiac comes from the Greek *zodiakos kyklos*, which means "circle of animals." All zodiac constellations are named after living creatures, with the exception of Libra the Scales.

The tilt of the earth's axis of rotation with respect to the space direction of the earth's revolutionary axis is 23.5 degrees. This is seen on the celestial sphere as a 23.5-degree angle between the ecliptic and the celestial equator. The celestial equator divides the celestial sphere in the same way the terrestrial equator divides the earth. It is a great circle projection of the terrestrial equator on to the celestial sphere that is 90 degrees away from the north and south celestial poles everywhere. Observationally, the celestial equator extends from the due east point on the horizon to the due west point on the horizon, reaching its maximum altitude when due south.

Because the ecliptic is inclined to the celestial equator, the sun will appear to cross the celestial equator two times each year and reach maximum angular distances of 23.5 degrees above and below the celestial equator once each year.

The June or vernal equinox (green or spring equal-night) is the time and place of the sun's apparent crossing of the celestial equator as the sun moves from below to above the celestial equator. The vernal equinox takes place on or about March 21 each year. This is the time of the official beginning of the spring season.

The summer solstice (summer sun-stop) is the time and place of the sun's greatest angular distance (23.5 degrees) above the celestial equator. The summer solstice takes place on or about June 22 each year, and is the official beginning of the summer season.

The autumnal equinox (autumn equal-night) is the time and place of the sun's apparent crossing of the celestial equator as the sun moves from above to below the celestial equator. The autumnal equinox takes place on or about September 23 each year, and is the official beginning of the autumn season.

The winter solstice (winter sun-stop) is the time and place of the sun's greatest angular distance (23.5 degrees) below the celestial equator. The winter solstice takes place on or about December 22 each year. This is the time of the official beginning of the winter season.

The sun's apparent ecliptic motion will systematically change the noon-time altitude of the sun. It will increase its angular height above the southern horizon from the time of the winter solstice until the time of the summer solstice, and decrease its angular height from the time of the summer solstice until the time of the winter solstice. The total variation between the two solstice (sun-stop) points is 47 degrees.

The position on the horizon where the sun will appear to rise and set will also systematically vary due to the sun's apparent ecliptic motion. The angular displacement between the summer solstice sunrise position and the winter solstice sunrise position is determined by geometric latitude. The sunset position varies in the same manner as the sunrise position.

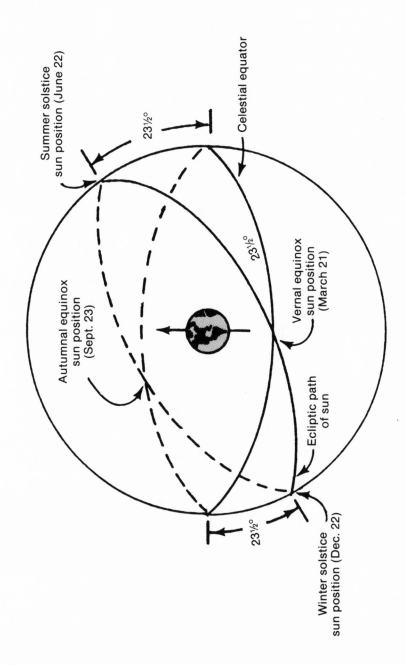

Summer solstice
sun position (June 22)

23½°

Celestial equator

23½°

Autumnal equinox
sun position
(Sept. 23)

Vernal equinox
sun position
(March 21)

Ecliptic path
of sun

23½°

Winter solstice
sun position (Dec. 22)

Sun positions on the celestial sphere.

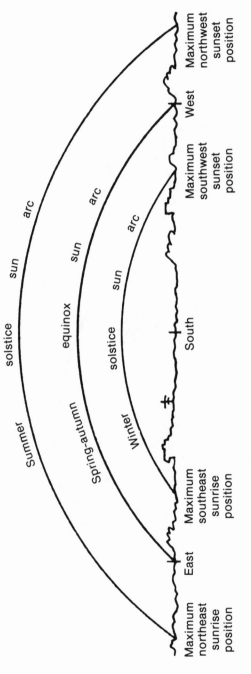

Sun arc and length of day.

The minimum angular displacement takes place at the terrestrial equator and increases with latitude. Eventually, a latitude is reached where the solstice sunrise position meets the solstice sunset position in the due north and south directions. At the time of the summer solstice, the sun will remain *above* the horizon for twenty-four hours. At the time of the winter solstice, the sun will remain *below* the horizon for twenty-four hours. At higher latitudes, the sun will remain above and below the horizon for periods of time greater than twenty-four hours. Those latitudes where the sun is visible for a twenty-four-hour period are referred to as the "lands of the midnight sun."

The land of the midnight sun begins at the Arctic Circle, latitude 66.5 degrees. Greek travelers knew that as they moved farther north the figures of the Big and Little Bear would rise higher in the sky. The Greek word for bear was *arkto,* so northern regions where the Big and Little Bear were highly visible were called the *arktikos* regions. We still call these regions the Arctic.

For mid-latitudes in the Northern Hemisphere the length of the day will exhibit fewer extremes. At the time of the winter solstice, when the sunrise and sunset positions are at their maximum angular displacement south of east and west, the sun will be visible above the horizon as it moves through a small arc. This will be the shortest day of the year, since for most of the day the sun will be below the horizon. At the time of the equinoxes, the sun rises due east and sets due west. The sun will be above the horizon for half of the day and below the horizon for half of the day. Equinox means "equal night." At the time of the summer solstice, when the sunrise and sunset positions are at their maximum angular displacement north of east and west, the sun will be visible above the horizon as it moves through a larger arc. This will be the longest day of the year, since for most of this day the sun will be above the horizon.

The length of the day increases from the time of the winter solstice to the time of the summer solstice, and decreases from the time of the summer solstice to the time of the winter solstice.

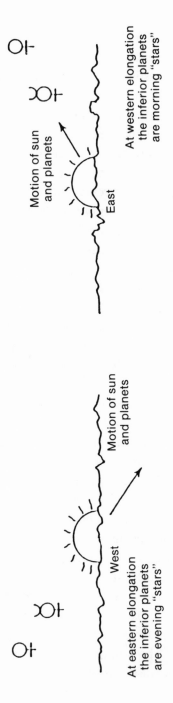

Morning and evening "stars."

The moon is the most recognized object in the night sky. It is the earth's only natural satellite and its closest neighbor in space. Its apparent size and brightness alone make it one of the prime attractions of the universe we view beyond the earth. Watching the moon change as it moves through the more distant stars can almost become a compulsion to stargazers, especially since the moon presents the most complicated phenomenon to the stargazer.

The most conscious observation of the moon is that its appearance changes from night to night. These changes in the proportion and shape of the lighted part of the moon, as seen from the earth, are called the phases of the moon.

The standard earth-moon-sun diagram explains the different positions of the new, quarters, and full moon in relation to the position of the sun and earth, but it has limited value for an understanding of the phases of the moon as seen in the true sky. An awareness and ability to correctly interpret the phases of the moon and all lunar-associated phenomenon, a lunar visual literacy, is best obtained by observing the moon from night to night.

On the night of the new moon, and for a few days after, there is nothing to observe. The shadowed side of the moon is facing the earth. But as the moon moves counterclockwise around the earth, more and more of the lighted side of the moon can be seen from the earth. This is the waxing, or growing, crescent moon. Observations of the waxing crescent moon, made soon after sunset over several nights, will show that the moon moves west to east, away from the sun. A closer inspection of the position of the moon, with respect to the background stars, will show that the moon also moves west to east through the background stars from night to night. The terminator, the line between the lighted and shadowed parts of the moon, will also be seen to move west to east across the lunar surface.

On the night of the first quarter moon, the terminator will appear as a straight line, and the moon will be 90 degrees to the east of the setting sun. The moon will have completed the first quarter of its orbit around the earth. The name of this phase is derived from the position of the moon rather than its appearance.

During the next week, the lighted part of the moon will continue to grow. But it will no longer be seen as a crescent. More than 50 percent of the side of the moon facing the earth will be illuminated. This will be the lunar waxing gibbous phase.

The moon will continue to move west to east away from the sun and through the background stars. It will eventually be far enough away from the sun to be seen in the eastern sky during late afternoon. It will also be easier to find in the night sky because it will set later each night.

The phenomenon of the full moon occurs when the moon and sun are on opposite sides of the earth. A fully illuminated moon rises as the sun sets, and sets as the sun rises. This is the moon phase that rises with an orange color, casts shadows, and turns the night sky blue. This phase receives its name from its appearance rather than its orbital position.

After the full moon the phases will be repeated in reverse order. The moon will continue to move eastward, but now it will appear as a waning, or paling, gibbous moon until it reaches third or last quarter. At this position in its orbit, the moon will have moved 270 degrees counterclockwise around the earth since new moon. It will be 90 degrees to the west of the sun, and will be seen from the earth as a half-illuminated hemisphere. Since the terminator and the night part of the moon also move from west to east across the lunar disk, the eastern half of the moon will now be illuminated.

The third quarter moon will be in the south around the time of sunrise since it is 90 degrees to the west of the sun. At successive sunrises, the position of the moon will be closer to the eastern horizon and the moon will appear as a waning crescent. The early morning crescent will get thinner and thinner, and finally be too close to the sun to be seen just before the moon is in conjunction with the sun once again.

The entire cycle from new moon to new moon, or full moon to full moon, takes 29.5 days. This is the time it takes the moon to gain one lap on the sun. It is defined as the moon's synodic (coming together) period. In general terms, a synodic period is the time between repeating configurations of celestial bodies with respect to one another.

The moon's nightly motion with respect to the stars is easily discernible to a lunar observer since the moon appears to move through an arc equal to its own mean diameter (one-half) in roughly one hour. Since the angular rate at which the moon moves along its orbit is much faster than the angular rate that the sun moves along the ecliptic, the moon will complete one revolution with respect to the stars before it catches up to the sun. The lunar sidereal (star) period of 27.32 days is the time it takes the moon to complete one revolution with respect to the background stars. In general terms, a sidereal period is the time it takes a celestial object to move 360 degrees with respect to the stars.

Observation of the moon through even a part of its synodic period will soon reveal that it orbits the earth with the same side always facing the earth. During the course of the month we always see some part of the same side of the moon. This is because the lunar rotation period is equal to the lunar period of revolution. Five billion years ago, the moon was much closer to the earth and rotating much more rapidly. It was largely the molten rock within the moon that was affected by the gravitational pull of the earth. The molten rock reacted to the gravitational pull of the earth in the same way the oceans of the earth are affected today by the gravitational pull of the moon. Lunar tides of liquid rock produced friction, and the friction caused the moon to slow its rate of rotation and move away from the earth. The moon eventually slowed down enough to become locked into the synchronous rotation circumstance, called tidal coupling, that we observe today.

The motion of the moon is similar, but more complicated, than that of the sun. The moon rises and sets each day due to the rotation of the earth, and it also moves eastward through the constellations of the zodiac. The motion of the moon as seen projected against the background stars of the celestial sphere is a great circle that is inclined five degrees with respect to the ecliptic.

The moon will display the same basic variation in moonrise and moonset points as the sun, and the same variation in the altitude of its meridian crossings as the sun. The five-degree difference will expand and contract the limits of the sun. The faster motion of the moon through the background stars will turn seasonal variations into monthly variations. The moon's behavior becomes complicated because the variation in moonrise and moonset points will themselves vary due to an 18.6-year periodic change in the orientation of the lunar orbit with respect to the ecliptic.

The lunar orbit crosses the ecliptic at two diametrically opposite points called the ascending node and the descending node. A perturbation of the moon's orbit by the sun causes these two crossing points to slowly move westward along the ecliptic. The nodes complete one 360-degree westward movement around the ecliptic in 18.6 years. This motion of the nodal points is referred to as the "regression of nodes." The full-moon rise and set points along the horizon will reach maximum and minimum angles north and south of due east and west in 18.6-year cycles due to the regression of nodes. Other lunar observations are tied to the regression of nodes cycle also.

When the moon and sun are both near or at a node, a solar or lunar eclipse takes place. An eclipse is a phenomenon that occurs when one celestial body becomes positioned such that it conceals or blocks another celestial body from some viewing place. A solar eclipse takes place when some part of the moon passes in front of the sun as seen from the earth. A lunar eclipse takes place when some part of the moon passes into the shadow of the earth caused by the sun.

Any opaque spherical body, such as the earth, will cast two types of shadows when it intercepts the light from a source of visible finite angular size, such as the sun. The darkest shadow, called the umbra, or shadow cone, represents a region that is completely devoid of the direct rays of the sun. No part of the sun is visible to an observer within an umbra. The partial shadow region, called the penumbra, is a transition region that receives direct rays of light from only some parts of the sun. Only part of the sun is therefore visible to an observer within a penumbra. An umbra and penumbra are formed by both the earth and the moon as they intercept the light from the sun.

In order for the moon to pass in front of the sun and cause a solar eclipse it must be in the new phase at or near the same ascending or descending node position in its orbit. Since a new moon occurs every 29.5 days, it might be expected that an eclipse would take place every 29.5 days. This is not the case because of the five-degree inclination of the orbit of the moon with respect to the ecliptic. The moon can be up to five degrees higher or lower than the sun at the time of the new moon. Five degrees is equivalent to ten lunar diameters.

The moon moves eastward in its orbit at approximately 2,300 miles per hour. The umbra, or shadow cone, of the moon will move through space in the same eastward direction and at the same rate. The thin line of places on the earth that intercepts the shadow cone is called the path of solar eclipse totality. These are the areas from which a total eclipse can be observed. The width of the path of totality varies as the distance between the moon and earth varies because of the moon's elliptical orbit. The maximum width of the path of totality is nearly 170 miles.

The statistical expectancy for a total solar eclipse to be visible at a given location on earth is once every 360 years. The maximum duration of a total solar eclipse is 7 minutes, 31 seconds.

The penumbra will cross the earth with a path width of approximately four thousand miles. Observers within the region will see the sun only partially covered by the moon. They will observe a partial solar eclipse.

The observation of a total solar eclipse is an emotional as well as a visual experience. It is well worth the time, effort, and expense to witness and experience one of nature's major league events.

A total solar eclipse begins when the moon first starts to move in front of the sun. This "first contact" appears as a small indentation in the western edge of the solar disk. The partial phase continues as more and more of the sun is covered by the eastward moving moon. The sky will gradually darken and colors will take on a different and eerie hue. Light traveling between the leaves of a tree will form "pin hole" camera projections of the partially eclipsed sun on the ground. Seconds before "second contact," the beginning of totality, shadow bands can often be seen racing along the ground in front of the eclipse shadow. Birds, insects, and plants will react as if night were falling.

Second contact occurs a little more than an hour after first contact. Within a few seconds several spectacular, short-lived visual phenomena occur. Just before the final covering of the sun, the "diamond ring" appears. The last part of the sun to be covered forms the sparkling diamond, while light shining around the edge of the moon forms a ring. Baily's beads, named after the English astronomer Francis Baily (1774–1844), will appear next. They appear as irregularly spaced beads on a thin thread. The beads are due to the light of the sun shining through the serrated profile of the moon that is formed by the mountains and valleys present at the edge of the moon. The appearance of Baily's beads will be different for different eclipses. During the time of totality, planets and first magnitude stars can often be seen in the proximity of the eclipsed sun. While the sky above will be dark, the sky along the horizon will be of a twilight darkness.

The most enduring sight of a total solar eclipse is the corona. This is a faint outer atmosphere of rarefied ionized gas that forms a halo extending irregularly around the eclipsed sun. The corona is visible to the naked eye only during a total solar eclipse. The shape of the corona is related to solar sunspot activity. Sunspots are dark markings that appear on the solar surface. The corona will appear uniform at times of maximum sunspot activity and greatly extended from the solar equator at times of minimum sunspot activity. Solar prominences can often be seen at the base of the corona. Prominences are gaseous eruptions that rise several hundred thousand miles above the surface of the sun.

Totality ends with "third contact," the point at which the moon begins to uncover the solar disk. The events following third contact duplicate the events preceding the time of second contact. Baily's beads, the diamond ring, and a repeat, in reverse order, of the partial phases of the eclipse will occur.

The fact that the phenomena associated with a total solar eclipse are visible is due to a remarkable coincidence. The sun, 864,000 miles in diameter, is approximately four hundred times as large as the moon, which has a diameter of 2,160 miles. The sun is also approximately four hundred times as far away. This coincidence causes the sun and moon to subtend nearly the same one-half-degree angle in the sky. It is this closeness in apparent size that allows Baily's beads, the diamond ring, prominences, and the corona to be seen during a total solar eclipse.

A partial solar eclipse begins and ends like a total solar eclipse, but it does not create the spectacular sights visible between the second and third contact of a total eclipse. Unless a partial solar eclipse is very close to being a total solar eclipse, it will go unnoticed.

An eclipse of the moon takes place when the moon moves into the shadow of the earth. This requires that the moon be at opposition, 180 degrees away from the sun, and therefore at the full phase. A lunar eclipse will be seen by an observer any place on earth where the moon is visible at the time of the eclipse. The chances of seeing a lunar eclipse are much greater than the chances of seeing a solar eclipse, even though lunar eclipses occur less frequently.

The type of lunar eclipse seen depends upon the path that the moon follows in moving through the earth's shadow. If the moon crosses the penumbral shadow and moves completely into the earth's umbral shadow, a total lunar eclipse will occur. If the moon crosses the penumbra, but does not completely move into the area of the umbra, a partial lunar eclipse will occur. If the moon moves into just the partial shadow of the penumbra or only part of the penumbra, but does not move into the umbra, a penumbral lunar eclipse will take place.

Lunar eclipses do not take place every full moon for the same reason that solar eclipses do not take place every new moon. The five-degree inclination of the lunar orbit to the ecliptic allows the full moon to pass above or below the umbra and penumbra at the time of full moon. A lunar eclipse can occur only when the full moon is at or near the ascending or descending node position on its orbit. The sun will be at the opposite node.

The duration of a total lunar eclipse is determined by the diameter of the umbra at the distance of the moon and the angle between the center of the umbra and the center of the moon. Under the best conditions, the maximum duration of the total phase of the eclipse will be approximately 1 hour and 40 minutes.

A total lunar eclipse does not offer the visual excitement of a total solar eclipse, but it is still well worth the effort to observe. The eclipse will begin when the moon inconspicuously moves into the penumbral shadow of the earth. The outer zone of the penumbral shadow will not darken the lunar surface by an amount that can be noticed by the naked eye. It will only be within a period of twenty minutes before first contact that a darkening of the eastern limb of the moon will be noticeable. At the time of first contact, the umbral shadow of the earth will touch the eastern edge of the moon. The moon will move into the umbra for approximately an hour as the circular-shaped shadow is seen to move from east to west across the lunar surface. This is the partial phase of the eclipse.

The partial phase of a total lunar eclipse will end at second contact when the western edge of the moon makes contact with the umbra. Totality will last while the moon remains completely within the umbral shadow. The duration of totality will depend upon how close the moon approaches the center of the umbra. The moon will not completely disappear during a total lunar eclipse. It will still be seen, but it will appear a reddish or coppery color due to light that is refracted by the lower atmosphere of the earth along the line between day and night.

Following third contact, the moon will start to move out of the umbral shadow and into the penumbral shadow. The partial phase events will be repeated in reverse order.

A new moon approaches the orbital node positions twice a year at roughly six-month intervals. These periods of time when solar eclipses are possible are called solar eclipse seasons. A full moon also approaches the orbital node positions twice a year at roughly six-month intervals. These periods of time when lunar eclipses are possible are called lunar eclipse seasons.

The regression of nodes causes the nodal points to move westward along the ecliptic in a period of 18.6 years. This causes the eclipse seasons to occur earlier each year by nineteen days, and solar and lunar eclipse patterns to repeat at 18.6-year intervals. This knowledge of the Saros, meaning "repetitious" period, allowed early astronomers to predict eclipses many years into the future.

The second-most conspicuous celestial objects in the night sky are the planets. There are five naked-eye planets that have been observed and studied by stargazers since the dawn of civilization. All five planets were discovered in the time before recorded history.

The planets look like stars, bright stars, but of all the stars visible in the night sky, only these five stars refused to remain "fixed" in position with respect to the other stars. These stars moved and changed in brightness. They did strange things. Attention understandably focused upon them.

Phases of the Moon

The phases of the moon as seen from the earth

1. New Moon
2. Waxing Crescent
3. First Quarter
4. Waxing Gibbous
5. Full Moon
6. Waning Gibbous
7. Third Quarter
8. Waning Crescent

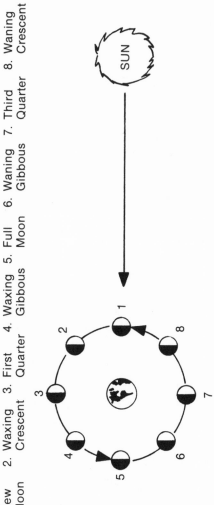

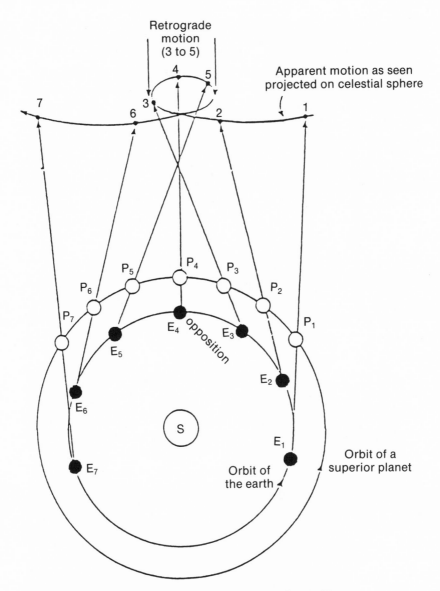

Circumstances of an apparent retrograde motion.

They were called the "wanderers." The word planet is derived from *planetes,* the Greek word for wanderer. The development of astronomy as a science arose from the desire to explain the motions of the five naked-eye planets. The individual naked-eye planets are known today, in their increasing distances from the sun, as Mercury, Venus, Mars, Jupiter, and Saturn.

Mercury and Venus are the inner planets. As seen from the earth, Mercury and Venus are always found within specific angles from the sun because they revolve around the sun in smaller orbits than the earth. Their evening and morning positions with respect to the horizon follow the horizon movement of the sun. This apparent domination by the sun led to their being classified long ago as the inferior planets.

The distances of planets are more conveniently measured in "astronomical units" than miles. It reduces the size of the numbers involved. An AU is equal to the 92,957,000-mile mean distance of the earth from the sun. The mean distance of Mercury from the sun is equal to 0.387 AU, and the mean distance of Venus is equal to 0.723 AU.

The angular displacement of Mercury and Venus from the sun as seen from the earth is referred to as an elongation. The amount of this elongation is dependent upon the size of the planet's orbit. The maximum elongation of Mercury is 28 degrees, while the maximum elongation of Venus is 48 degrees. The two planets will be seen at their maximum elongations when they are in a position in their orbit that intersects the tangent to the orbit as seen from the earth.

Mercury's orbit has a high ellipticity and therefore a distance from the sun that varies considerably from its mean distance. It ranges from 0.30 AU at perihelion ("close-sun") to 0.47 AU at aphelion ("far-sun"). The angular value will vary between these extremes depending upon the position of Mercury in its orbit when a maximum elongation takes place. The maximum elongation will be 18 degrees with Mercury at perihelion and 28 degrees if it occurs with Mercury at aphelion.

The orbit of Venus is larger and more nearly circular than the orbit of Mercury. Venus's maximum elongations are the same 48 degrees from one maximum elongation to the next.

Two different elongations with respect to the sun are possible for Mercury and Venus: eastern elongation and western elongation. The inner planets can be seen to the east or to the west of the sun. In going from an eastern to a western elongation, the counterclockwise orbiting Mercury and Venus must pass between the earth and sun. When an inner planet passes between the earth and sun, it passes through inferior conjunction. In going from a western to an eastern elongation, Mercury and Venus must pass behind the sun as seen from the earth. When an inner planet passes behind the sun, it passes through superior conjunction.

The east to west movement from maximum eastern elongation to maximum western elongation requires a lesser orbital movement of Mercury and Venus than the maximum western to maximum eastern elongation movement. This is because of Mercury and Venus's closer proximity to the earth at inferior conjunction. For the same reason, the maximum eastern to maximum western elongation movement requires a lesser time than the maximum western to maximum eastern elongation movement. The maximum elongation to maximum elongation times for Mercury average 44 days and 72 days. The times for Venus average 144 days and 440 days.

An inner planet at any eastern elongation will set after the sun and rise after the sun. It will be seen in the western sky as an "evening star," but it will not be visible in the morning sky because of the brightness of the already risen sun. After the time of a maximum eastern elongation, the planet will move at an accelerated pace into the glare of the sun and will quickly reappear at a later date on the other side of the sun. It will continue to move at an accelerated pace to its maximum western elongation.

An inner planet at any western elongation will set before the sun and rise before the sun. It will be seen in the eastern sky as a "morning star," but it will not be visible in the evening sky because of the brightness of the still visible sun. After the time of a maximum western elongation, the planet will move at a much slower pace back into the glare of the sun. It will reappear at a later date on the other side of the sun and slowly move back to a maximum eastern elongation position.

The visibility of Mercury or Venus depends simply upon the amount of time it will be in the morning or evening sky. Brightness is not a problem. The time of visibility depends upon the elongation angle of the planet and the angle that the ecliptic makes with the horizon. The ecliptic angle with the horizon is important since Mercury and Venus follow the sun's apparent motion along the ecliptic as they move from one side of the sun to the other side of the sun.

The brilliancy of both planets is a factor of their distance from the earth and the amount of light reflecting area facing the earth. The inner planets go through the same cycle of phases as the moon, but the phases are only visible with a telescope. For Venus, the two brightness factors provide the condition of maximum brilliancy when the planet is approximately thirty-six days to the east and west of inferior conjunction. Venus can attain an apparent brightness that is fifteen times greater than the brightness of Sirius, the brightest star visible. For Mercury, maximum brilliancy occurs a few days before and after superior conjunction when Mercury is lost to the naked eye in the glare of the sun. It is brighter than any star, with the exception of Sirius, during the times of eastern and western elongation.

Venus was probably the first "wandering star" discovered by ancient stargazers because of its brightness in the early evening and early morning skies, and its rapid movement in and out of the glare of the sun. Mercury was probably the last to be discovered, and remains the planet that has been seen by the fewest number of people because of its low horizon angle and short time of visibility.

Mars, Jupiter, and Saturn are not dominated by the position of the sun and were therefore designated in ancient times as superior planets. They have orbits that are larger than the orbit of the earth, with mean distances from the sun of 1.52 AU for Mars, 5.20 AU for Jupiter, and 9.54 AU for Saturn. As seen from earth, the superior planets move from west to east with respect to the background stars along paths that are slightly inclined to the ecliptic.

At the position of opposition with the sun, a superior planet will be closest to the earth and appear at maximum brightness. It will be seen in a part of the sky that is 180 degrees away (opposite) from the sun. The planet will rise at sunset and remain visible all night. At maximum brightness, Mars and Jupiter outshine Sirius, while Saturn at maximum brightness only rivals Sirius.

The reddish-colored Mars is the fastest moving of the outer naked-eye planets. It travels from west to east along the ecliptic on an average of nearly one-half degree per day. The average synodic period of Mars, as measured from opposition to opposition, is 780 days. The sidereal period of Mars is 687 days. This is the time it takes for Mars to complete a 360-degree counterclockwise revolution of the sun with respect to a space reference system.

Jupiter is the next fastest moving of the superior planets. It takes Jupiter a sidereal period of 11.86 years to move completely around the ecliptic. Jupiter moves from one zodiac constellation to the next each year. Because of its rather slow motion, the earth gains one revolution on Jupiter in a little more than one year, giving the planet a synodic period from opposition to opposition of 399 days.

The planet Saturn is the slowest of the naked-eye planets. It takes a sidereal period of 29.46 years to move completely around the ecliptic. Saturn moves from one zodiac constellation to the next every 2.5 years. The earth gains one revolution on Saturn in one year and thirteen days, giving the planet a synodic period from opposition to opposition of 378 days.

The most perplexing motion of the outer planets is that of

"retrograde" motion. For the greater part of the year the outer planets move from west to east through the background stars in what is termed "direct" motion. As the time of opposition approaches, however, the outer planets will appear to slow down in their direct motion. The direct motion will cease altogether at the first stationary point. The planets will then begin their "retrograde" motion from east to west through the background stars.

Retrograde motion is a backward motion contrary to the usual motion of the planets. The retrograde motion will increase to a maximum at opposition. It will then decrease and finally cease altogether at the second stationary point. After reaching the second stationary point, the planets will move in direct motion from west to east through the background stars until the approach of the next opposition.

The retrograde motion of an outer planet is not a real motion of the planet. It is only an apparent motion seen from earth as earth overtakes and passes the slower moving outer planets. As seen from a point above the solar system, the earth and outer retrograding planet continue to move in the same counterclockwise direction during the entire retrograde episode.

The duration and angular size of the retrograde motions of the individual outer planets vary because of a combination of their distance from the earth at opposition and their orbital velocities. Mars appears to move in retrograde for a period of approximately two months. During that time it moves through an angle of nearly 15 degrees. Jupiter appears to move in retrograde for a period of approximately four months. During that time it appears to move through an angle of nearly 10 degrees. The retrograde angle for Jupiter is less than Mars, partly because Jupiter, at the time of opposition, is eight times as far away from earth as Mars. Saturn also appears to move in retrograde for a period of approximately four months. During that time it appears to move through an angle of nearly 7 degrees. The retrograde angle for Saturn is less than Jupiter partly because Saturn, at the time of opposition, is twice as far away from earth as Jupiter.

Most people are unaware of the retrograde motions of the outer planets. It doesn't come up in conversation very often. It is only when attention is focused on a particular celestial phenomenon or object that things are seen that would normally have been missed or disregarded. Awareness is the prerequisite to observation. And observation is the prerequisite to true understanding. There are many rare phenomena, as well as common everyday phenomena, that can be observed by the more aware celestial observer.

The rare sight of a bright naked-eye comet is an attention-getter. It may even come up in conversation. The comet will appear with a star-like head and extensive tail stretching away from the sun. The word *comet* means "long haired," and it very aptly describes the comet's appearance in the night sky. Comets bright enough to be seen by unaided vision appear on an average of every six or seven years. On an average of every thirty or more years, an extremely bright comet will appear.

Deep in space a comet is a dirty snowball of small particles held together by frozen gases. As the comet's dirty snowball nucleus approaches the sun, its frozen gases are vaporized by solar radiation to form a spherical coma measuring thousands of miles in diameter. The solar wind of protons and electrons emitted by the sun forces the coma material to form the comet's tail. The tail always faces away from the sun. This tail, or "solar wind sock," can grow to lengths approaching one hundred million miles. The comet's brightness from reflected sunlight increases as it approaches the sun. It reaches maximum brightness near the perihelion position of its elliptical orbit, and then fades as it moves away from the sun.

Occasionally the aware stargazer will see a sporadic meteor streak through the dark night sky. Sporadic meteors are small sand-grain-sized fragments of stone and metal that strike the earth's atmosphere at random times and in random directions. The fragments are heated and vaporized by friction with the atmosphere to produce momentary, luminous trails visible to the naked eye. Larger meteors sometimes form trails that will exist for several minutes and even become distorted by air currents in the upper atmosphere. Under the best of conditions, an observer can expect to see from five to ten sporadic meteors per hour. This represents an earth-wide total number of sporadic meteors measuring in the millions per day. It also represents a total mass of several tons.

Sometimes a sporadic meteor is large enough to survive atmospheric friction and reach the earth. It is then called a meteorite. Some meteorites have been able to produce sizeable craters.

On predictable dates during the year, meteors are seen to radiate with greatly increased frequency from a particular part of the sky. These meteor showers result from the passage of the earth through the orbital path of a comet. A comet gradually disintegrates and distributes its grain-sized stone and metal debris throughout its orbit with each perihelion passage. When the earth intercepts the cometary debris while passing through the region of the comet orbit, large numbers of meteors appear to come from a single "radiant" point in the sky.

The apparent radiant point is a perspective effect. The meteor particles are really traveling though space on parallel paths. As seen from the earth, shower meteors seem to diverge from a point in the same way as railroad tracks, tunnels, and telephone poles can be seen to diverge from a point. The debris from Comet 1866 I produces the annual fifty-per-hour Leonid meteor shower every November, and a spectacular shower every thirty-three years. On the night of November 17, 1966, it produced a twenty-minute shower of 2,400 meteors per minute!

Circling the dark sky away from the lights of urban areas is a broad diffuse band of light called the Milky Way. It extends from its brightest area in Sagittarius into Aquila, through Cygnus, Cepheus, and Cassiopeia. It continues between Orion and Gemini to Canis Major, Vela, Crux, and then back to Scorpius. A closer inspection will show many localized features within the Milky Way band. For instance, the brightness and star density vary greatly. In its brightest area, the constellation Sagittarius, the star density is ten times per square degree what it is in the faint area of Auriga.

A feature visible in the late summer and early fall Milky Way is the Great Rift. The Great Rift, which is a dark area that splits the Milky Way into two segments between Cygnus and Scorpius, was once thought to be a hole through the Milky Way. Actually, it is a huge obscuring gas cloud or nebula.

The Milky Way consists of the estimated hundreds of billions of unresolved stars that constitute the Milky Way spiral galaxy. The stars appear as a band of diffuse light because they are too far away to be separated into individual stars by the unaided eye. The light appears brightest in the direction of Sagittarius because that is the direction of the center of the galaxy and the greatest number of stars.

There are other galaxies besides our Milky Way galaxy. The Andromeda galaxy is a Milky Way-sized galaxy that appears to the naked eye in a dark sky environment as a small smudge of light. At a distance of 2.2 million light years, it is the farthest object in the universe visible to the naked eye. The larger appearing and brighter Large and Small Magellanic Clouds seen from the Southern Hemisphere are two irregular dwarf galaxies held by gravity to the Milky Way.

If the stars of the Milky Way were evenly distributed among the population of the United States, everyone would receive a thousand stars each. The combined stars of any five people would represent the total number of stars visible to the naked eye. And for each star distributed, there is a galaxy in space.

An attempt to "save the phenomena" took a simulation approach in the twentieth century. On October 21, 1923, an electro-mechanical, optical creation opened its conical eyes and a new universe was born. The interior surface of a 32-foot dome came alive with 4,500 stars. The night sky had been captured and brought indoors to an environment free of the constraints weather and time have traditionally imposed upon stargazers. The world's first public planetarium program had taken place in the Deutsches Museum in Munich, Germany.

The projection planetarium was the Mark I, the invention of Dr. Walther Bauersfeld (1879–1959) of the Zeiss Optical Company. Under his guidance improvements were made in the design of the original single starball projector.

The Zeiss Mark II featured two hemispheres, one for the northern sky stars and one for the southern sky stars. The mechanical planet cage was then placed at the center of the two-hemisphere "dumbbell" planetarium instrument. Projectors for the meridian, celestial equator, and ecliptic were added. The major improvement was that the new model allowed an audience to view the sky from anywhere on earth.

The first Zeiss planetarium came to the United States in 1930 when philanthropist Max Adler donated a Mark II to the city of Chicago. This move was countered by Adler's friend, Samuel Fels of Philadelphia. Fels presented a Zeiss Mark II planetarium to the new Franklin Institute that was to open in early 1934. It was the impact of the Fels Planetarium on a staff member that spawned the development of a new model star projector and the founding of the first American planetarium manufacturer, Spitz Laboratories, Inc.

Armand Spitz (1904–1971) was Director of Education at the Fels Planetarium of the Franklin Institute in the 1940s when he decided the planetarium was "the greatest single teaching instrument ever invented." His dream was to produce an inexpensive star projector that would make planetariums available to smaller cities, schools, and colleges. Armand Spitz became the "Henry Ford" of planetariums.

Spitz solved the cost problem with a starball of twelve black, perforated plastic pentagons forming a dodecahedron. It was a simulation device designed for teaching astronomy. The resulting starfield, from light shining through small holes, did not produce a realistic illusion, but five hundred Spitz Model A planetariums could be purchased in 1947 for the same cost as a Zeiss planetarium instrument. The planetarium became accessible to schools, colleges, and smaller museums. The Spitz Model A was followed by a succession of improved and more complex optical lens models. These models eventually achieved a realistic view of the night sky.

The principle of planetarium design remained constant for half a century. A star projector, or pair of star projectors, was attached to a clockwork planet projection mechanism. The planet projectors were driven in fixed motion patterns by gear systems called analogs. This allowed the operator to demonstrate the annual motion of the planets against the starfield as seen from earth.

The new challenge for planetarium manufacturers was to design systems that could save the phenomena seen not only from earth, but from any point in the solar system or from the stars beyond. The traditional planet analog approach had to be abandoned and new technologies had to be introduced to the industry.

The first solution to the challenge was found by Spitz in the rapidly growing computer industry. The planets were designed as discrete projectors with no mechanical couplings to the star projector. The motions of all the projectors were tied together electronically through the use of computers. Any relationship between celestial bodies that could be calculated by a computer system could now be projected on the dome.

The Spitz Space Voyager can present the sky as seen from any position on the surface of the earth, or any position on the surface of any body in the solar system. It can also present the motion of the stars, moon, sun, and planets as they would be seen from any planet or natural satellite as the body rotates on its axis, revolves around the sun, or as the observer moves about the surface of the body. It also can move an audience in simulated interplanetary flight through an orbit at any angle of inclination about the sun, or through the orbit of any planet or satellite in the solar system.

Evans and Sutherland, a leading American computer graphics company, saved the phenomena with computers in yet another way. The Digistar planetarium uses a fisheye lens to project a video night sky onto the planetarium dome. The strength of the Digistar lies in its ability to move through vast reaches of space and time. The audience is capable of moving at hypervelocity speeds to see the universe from the perspective of any star in its data base. The computational ability of the Digistar's computers also allows the audience to see the changes in the night sky and constellations over vast amounts of time due to the gradual motions of stars through space.

Celestial phenomena have been saved in still another way— by mimicry. The sign language used by deaf people in the United States contains many accurate visual representations of naked-eye astronomy.

Sunrise, for example, is expressed in sign language by holding your left arm in front of your body and pointing right with your palm down. This is the horizon. Your right thumb and index finger form an "O." This is the sun. When the sun moves upward behind your left forearm, you have visually reproduced, from your perspective, both the direction and movement of the sun at the time of sunrise. You have "signed" sunrise to someone who understands sign language.

Using the same arm arrangement, the "sun" is moved down below your left forearm to produce sunset. This is not an accurate representation, since the sun really sets in the opposite direction of sunrise. It is just as easy, however, to understand the possible origin of the sunrise sign as it is to understand the reason for the convenient deviation from a true representation of appearances for the sunset sign.

Noon time, in sign language, is indicated by holding your right hand straight up with your right elbow on the fingertips of the horizon formed by your left arm. This is an accurate symbolic representation of the fact that the sun reaches its highest altitude each day at noon.

The sign for "all night" is another accurate picture of nature. Your left arm is again held in front of your body, palm down, pointing right. The fingertips of your left arm should be touching the crook of your right arm. Your right arm swings down and then toward your left to accurately demonstrate the nightly course of the sun from west to east, beneath the horizon.

The concept of the celestial sphere is used in sign language. A sweeping motion of your open right hand from left to right above eye level duplicates the east to west motion of the sky due to the earth's rotation. In sign language, this signifies "sky" or "heavens."

The sign for star is equally as expressive of the visual night sky. Point up with your two index fingers. Now, with your palms facing forward, move your right index finger along the side of your left index finger. Your two hands will move naturally toward the sky. Your two index fingers are striking each other as you would strike flint to produce twinkling of light. The twinkling sparks of light in the sky are what we call stars.

The manual signs have a history just as written and spoken words have a history, but the history of the manual signs has not been documented as thoroughly. It would be nice to say for sure that the signs for sunrise, noon, all night, the sky, and the moon were all meant to accurately duplicate the motions of the apparent naked-eye sky, but the statement may not be true simply because sign language is very definitely a right-handed language. The connection may be fallacious, and the motions may only be accurate because the language was developed to be "spoken" mainly by the movement of the right hand.

If Plato were alive today, he would probably be proud of the accomplishments of his students in saving the phenomena.

3

Magic Sky Astronomy

"Philosophy is written in this grand book, the Universe, which stands continually open to our gaze. But it cannot be understood unless one first learns to comprehend the language and interpret the characters in which it is written."

This was written by Galileo in the seventeenth century, but it could have been expressed millennia before. From earliest times people have looked into the sky in an attempt to learn about the universe in which they found themselves. But stargazers did not always read the sky in the modern sense. In the beginning, they created the sky. They named the stars. They created the constellations. They created the sky myths. They created the magic sky of primitive wonder and worship.

Myths concern gods and goddesses, and extraordinary heroes and heroines involved in extraordinary adventures. In relation to everyday life, they seem truly "out of this world." They appear to be the purely fictitious creations of fertile imaginations. They are not.

If we have been unable to find "heaven on earth," at least we have been able to find "earth in heaven." Sky myths reflect human history, human knowledge, and human concerns. Surprisingly, the more we learn today about our universe and ourselves, the more we often find fragments of this knowledge in the ancient myths. We also find relics of the era of the magic sky that have survived to play a part in our lives today.

The primary myth of a culture is the "in the beginning" myth. It sets the stage for the philosophical outlook of the culture. Sky creation myths can be classified as one of two types: those presenting cosmic rhythms that extend from undetermined time-past to indefinite time-future, and those that show creation as an ordered, sequential, nonrepeating event.

Eastern cultures are dominated by cyclical myths such as the recurring dance of the Hindu god Shiva. Western cultures are dominated by linear myths such as the ordered cosmos that evolved out of the disordered chaos to create the Olympian gods.

According to Hindu mythology, our physical universe began with a bit of terpsichore. It began with the dance of the four-armed god Shiva, the Lord of the Cosmic Dance and God of Creation and Destruction. Shiva's dance begins each new cosmic dream of the god Brahma.

The day of Brahma lasts 4,320 million earth years, and the night is equally as long. One year of Brahma's life has ended after 360 such days. The life of Brahma lasts one hundred years, and he is now in his fifty-first year.

After Brahma's lifetime another sleeping and waking god will appear. There was a similar god before Brahma. The Hindu universe is therefore a series of cycles that extends from undetermined time-past to indefinite time-future.

This ancient creation myth seems totally removed from our modern scientific concept of the creation of the universe. It appears to be one of pure fantasy dreamed up by a primitive nonscientific culture. But there are some interesting parallels between the Hindu creation myth and our present understanding of the cosmos.

For instance, Brahma's life is half over. So is the sun's life. From Brahma's dream, life on earth is possible. From the sun's energy, life on earth is possible. Brahma has dreamed other dreams, and the sun has been other stars.

When the sun evolved from a great gas cloud some five billion years ago, some of the atoms in the cloud had previously been part of other stars. The elements of the sun more complex than helium were created in the energy-producing thermonuclear furnace of another star's core or in its death throes. They were then distributed during the cataclysmic supernova events that terminated the life of this star.

Some of these atoms were in the nebula out of which the sun and solar system later formed. Some of these atoms are found in the complex organic molecules that produce the wide varieties of life found on earth.

Five billion years from now, as far into the future as the sun has existed in the past, the sun will evolve to become a red giant star, and the dream of life on earth will be over. The expanding sun will boil away all life on earth.

Life on earth will end, but the universe will continue. The universe began fifteen to twenty billion years ago with an explosion out of which resulted the expanding universe of this creation event.

The future of our universe is dependent upon the unknown total mass of the universe. If the mass within the universe is greater than a critical amount, gravitational forces will act as a restraining break on the expanding universe of today. The universe will eventually stop expanding. It will then proceed to contract and destroy itself in a cosmic crucible of unimaginable temperature. Our dream will be over.

A universe exploded from such a crucible in the distant past to create our present expanding universe. Would this not happen again in the future following a contraction of the universe? Would this end to our present universe not create another expanding universe? Has it happened in the past to create universes before the present universe? How many Brahmas dreamed in the past? How many will dream in the future?

The ancient Greeks saw the universe as having evolved from a disordered state of unformed matter that they called "chaos." This confused universe was reassembled into an orderly arrangement of stars and planets—a "cosmos." Out of this ordered array of matter came the Olympian gods. In this mythology, the gods were neither responsible for the universe nor superior to it. They were created by the universe, and the universe had but one beginning. It was an ordered, sequential, one-event universe.

The Olympian universe is a nontechnical version of the "Big Bang" universe of present research. The expanding universe is the cosmos that evolved from the chaos of an explosive creation event. The known mass of the universe, at the present time, is insufficient to halt the present universe's expansion. This being so, the universe will continue to expand forever. Stars will form, evolve, and finish their lifetimes as cold dark bodies. As the stars burn out, so the galaxies will fade and the universe will become an endless dark tomb void of life and light. It will exist, but it will be energy-dead.

Other myths were needed to explain the appearance and events of the night and day sky.

The Egyptian mythical explanation of the sky reads like a pharonic soap opera. It involved Nut, the sky goddess, and her son Osiris.

Nut was envisioned as the all-encompassing sky with stars decorating her body. Her arched body was supported by her hands and feet above her husband Geb, the god of the earth. At dawn, Nut gave birth to the fiery disk of the sun, which moved across her body until nightfall. It was then swallowed to pass through her body to be reborn the following morning.

In later Egyptian mythology, the sun was considered a god called Amon-Re, who navigated a celestial Nile in his royal barque. The summer flooding of this celestial Nile brought the sun higher in the sky and closer to earth. Its reduced winter flow moved the sun lower in the sky and farther away. This accounted for the seasons and the varying apparent paths of the sun across the sky during the different seasons.

Nut's son Osiris, a benevolent king of Egypt, was treacherously trapped in a coffin and thrown into the Nile by his jealous brother,

Set. Osiris's queen, Isis, found the castaway coffin and secretly returned Osiris's body to Egypt. But Set discovered Osiris's returned body and cut it into pieces to be hidden in different places throughout the country. Isis tracked down the individual pieces and wrapped them together to make the first mummy.

Isis then magically conceived Osiris's child, Horus, the falcon-headed god. (The magic came not from the fact that he was dead at the time, but from the fact that an important part of Osiris was never found.) When Horus reached adulthood, he avenged his murdered father by killing Set, losing an eye in the process. Horus then became the king of the living on earth, and Osiris ascended into the heavens to become the judge of the dead. The sun became the symbol of Horus's good eye and the moon the symbol of his lost eye. Every month Horus's eye was seen to be destroyed during the waning of the moon and healed during the waxing of the moon.

The pharaohs of ancient Egypt used the stars and the Osiris myth to attain immortality. The goal of the ancient Egyptian pharaohs in life was to ensure the existence of their spiritual *ka* for eternity in the afterlife. The pharaohs, as gods, were able to claim a position among the stars upon their deaths. The most desirable stars to ascend to were the "imperishable" circumpolar stars. These were the stars in the northern sky that never rise and never set. These are the stars that never died.

The pyramid burial tombs of the pharaohs were considered to be stairways to the stars. Entrance corridors to the north were used to bring the mummified pharaoh's physical body to its house of eternity. The same entrance, although filled and hidden after interment, was thought to provide an exit for the pharaoh's seven spirits to depart into the circumpolar sky.

Sometimes a more unencumbered route was provided. A shaft leading from the King's Chamber of the Great Pyramid of Khufu at Giza points in the direction of the north celestial pole, the point around which the north circumpolar stars move. This was the express route.

Magic sky myths are not entirely a thing of the past. They are the result of a primitive knowledge base, not time. Far from the space age of the twentieth century, and deep within the rain forest surrounding the tributaries that begin the 4,000-mile-long Amazon River, there lives a tribe of 450 people little known to the outside world. They are called the Waorani.

The Waorani live a simple life based upon their desire to survive within their jungle environment. They are nomadic hunters who stalk monkeys in the thick, jungle canopy with twenty-foot-long blow guns and poison darts. They have no leaders, no laws, and no history. But they do have a cosmology, a view of the structure of the universe. The dark night sky and the brighter stars are difficult to ignore even in the jungle. The structure of the Waorani universe closely resembles Aristotle's "onion" universe of concentric shells from two thousand years ago. They view the earth as beneath a succession of celestial levels. The sun and moon belong in the lowest level, the stars are in a higher level.

The Waorani are familiar with the movement of the sky but see no need for an explanation of why these objects move as they do. Celestial objects simply disappear below the western horizon and reappear above the eastern horizon. What happens between the disappearance and reappearance is not a matter that requires explanation.

The celestial Milky Way is the Tapir's Heaven. The tapir is a nocturnal mammal indigenous to the lush jungles of South America. Its heaven is the goal of the Waorani, but getting there has nothing to do with the good or evil doings of the Waorani on earth. In order to get to the Tapir's Heaven, the departed Waorani must attempt to swing on a vine over a dreaded giant Anaconda snake. Success leads to an afterlife in the Milky Way. Failure leads to reincarnation as a termite.

As fantastic as some of the sky myths appear to be, there is often an ancient element of knowledge hidden within them. The fantasy elements of myths often have a basis in reality.

The Gemini myth is certainly such a case. One form of the myth begins with Zeus playing the leading role as a "peeping Tom." The object of his attention was Leda, the beautiful queen of the king of ancient Sparta. She was taking her evening bath in a secluded lake. Zeus was anxious to join her, but social custom being what it was, he realized that his sudden and uninvited appearance would have a detrimental effect upon his intentions. He therefore transformed himself into a magnificent white swan and glided gracefully after the unsuspecting Leda. The disguise was a success.

Nine months later Leda gave birth to quadruplets. They were named Pollux, Castor, Helen, and Clytemnestra. Leda's husband, Tyndareus, should have been a little suspicious, since Pollux was born with the godly trait of immortality.

Helen was later abducted by Paris. She became the beautiful Helen of Troy, "the face that launched a thousand ships" and began the Trojan War. Her sister Clytemnestra married Agamemnon, the commander of the Greek forces that besieged Troy. And Pollux and Castor became the inseparable brothers whose most famous adventures occurred when they accompanied Jason and the Argonauts in search of the Golden Fleece.

Castor was eventually killed. Pollux then requested that Zeus take away his own immortality unless he could share it with his brother. Zeus never seemed to honor a request without some little change. In this case, he put the twins in the sky as a constellation that spends part of its time with the gods in the heavens and part of its time in the underworld, the land of the dead. He also added his seductive swan disguise to the sky as the constellation Cygnus the Swan.

The most preposterous aspect of this myth is not the event of the birth of quadruplets having two different fathers. That Pollux and Castor could be twins with different fathers is not a creation of fiction. It is based on a little known medical fact. The myth is based on an ancient knowledge of a rare phenomenon known as superfecundity, the fertilization of more than one ovum with one female menstrual cycle by two different males.

Sky myths have also been extended to the moon, and in particular to the dark and light areas of the moon. This is not surprising since the moon is the earth's only natural satellite and its closest neighbor in space. Its waxing and waning and movement through the stars of the night sky make it impossible to ignore. The myths of "something in the moon" are worldwide and as ancient as any other myth. They mostly serve to teach a lesson.

The dark areas seen on the lunar sphere were called *maria* (plural form of *mare*) or seas by Galileo after observing them with his telescope in the early seventeenth century. He made a mistake in doing this. It was a mistake that has never been corrected. The maria are not seas of liquid water, as thought by Galileo. They are the frozen seas of once flowing hot lava that rose through fractures in the lunar crust. Approximately 40 percent of the lunar surface visible from earth is covered by these broad, relatively smooth, circular plains measuring hundreds of miles in diameter.

The myths do not all agree. Western civilization saw a man in the moon; the Chinese saw a three-legged toad; the Mexicans saw a hare; and the Indians of northeastern North America saw a girl in the moon. Many of the myths do contain the commonality of the inhabitant being on the moon as a punishment for a bad deed, or as the reward for a good deed. These are lesson myths.

A Teutonic myth pictures the "man in the moon" carrying a bundle of sticks on his back. His sin was working on the Sabbath. A god appeared while he was busily engaged in cutting wood. When the man was reminded of the error of his labors, he foolishly laughed and remarked that Sunday and Monday were all the same to him. It was then declared that his punishment would involve standing in the moon for a perpetual Monday (moon-day).

In a Buddhist myth, the god Sakkria disguised himself as a starving man and went to several animals for help. The monkey provided him with mangoes and the fox with milk, but the hare could offer nothing but grass. The god suggested that the hare itself would make a tasty meal. The hare agreed. The god then built a fire and asked the hare to jump into it in order to save him the trouble of having to kill it. The hare agreed again, but as it leaped into the fire, the god caught it. Sakkria then decreed that the figure of a hare should remain forever visible on the surface of the moon as perpetual reminder of the hare's excellent willingness toward self-sacrifice.

Several moon myths explain the maria-formed figure as due to some action of the moon itself. These are not necessarily lesson myths.

The New Zealand aborigines, for instance, have a myth about a man who stumbled one night and sprained his ankle. His cries for help were so loud that the moon came down and tried to pick him up. The man was understandably terrified and grabbed on to a bush. The moon then pulled so hard that the man, the bush, and its roots were pulled up into the sky. According to the New Zealand aborigines, the man, the bush, and its roots can still be seen in the dark areas of the moon's surface.

Indians of British Columbia have a similar myth that identifies the moon "figure" as a crying child who was neglected by its mother. The child was crying so loudly for water one night the moon came to the door of its lodge with a pot of water. The child seized the water and was carried into the sky, where its face can still be seen in the pattern created by the dark and light areas of the moon.

Nursery rhymes are generally thought to be mindless little jingles that are taught to kids as part of the process of developing their language skills. The most commonly known nursery rhymes date back to seventeenth-century England. As ridiculous as their content appears to be, the origins of many nursery rhymes are embedded in aspects of political reality. It should therefore not be surprising to find there is one nursery rhyme that has its origins in a "man in the moon" myth. It is the familiar nursery rhyme called "Jack and Jill." The origin of Jack and Jill is not English, however; it is Scandinavian.

Originally Jack was Juki, and Jill was Bil. They were two young children who lived in the woods with their parents—a woodcutter and his wife. One night, just after the sun went down and the light from a full, white, spotless moon was flooding the landscape, Juki and Bil were sent out to "fetch a pail of water."

Just as Juki was about to toss the pail into the well, he noticed a beautiful reflection of the moon in the water. It was so beautiful he was reluctant to spoil the vision by dropping the bucket and disturbing the water. Eventually, duty outweighed aesthetics, and the pail and Juki "fell down" into the well. The rope attached to the bucket somehow caught Juki's shirt and pulled him into the deep, dark, well enclosure. He hit his head on the way down and "broke his crown." Juki called for help and his obliging sister Bil "came tumbling after" him.

The moon looked down at the two children thrashing around at the bottom of the well and remembered all the nice things the children had said about the beauty of reflected moonlight. So the moon sent down a moonbeam to pick up Juki and Bil, and the pole and bucket they had carried with them to the well.

The children's parents heard all the commotion and rushed to the well. As they looked down into the well, the water calmed and the image of the moon reappeared. This time, however, the image was not that of a white, spotless moon. The moon now had dark places that outlined the figures of two children carrying a pole and bucket.

There is a constellation in the early evening summer sky that serves to remind us of the close connection that once existed between the magic sky and medicine. It is the constellation Ophiuchus the Serpent Holder. The constellation is also known as Ophiuchus the Healer.

Snakes have been associated with healing powers since way back when. This is probably because snakes have the ability to shed their skins and thereby give the appearance of being restored to life or rejuvenated. Snakes are still used to symbolize the art of healing. The symbol of the AMA, the American Medical Association, is a staff with two snakes shimmying up it.

There was a time when astronomy, in the form of astrology, was a required part of all serious medical studies. The ancient Greek physician Hippocrates, who is recognized as the father of medicine and the supposed author of the Hippocratic oath of ethical behavior, is at least partly responsible for this. Hippocrates is supposed to have declared, "A physician without a knowledge of astrology has no right to call himself a physician." It was through the required study of astrology in his medical education that the Polish astronomer Nicholas Copernicus (1473–1543) was first introduced to astronomy.

The connection between astrology and medicine was based upon the belief that each of the parts of the body was influenced by the members of the solar system and the signs of the zodiac. The connections have been graphically illustrated over the centuries in the familiar farmer's almanac pictures depicting a disemboweled man circled by the signs of the zodiac.

When a patient of days past sought the help of a physician, the physician constructed an astrological chart, a picture of the heavens, for the time of the onset of the illness. This chart, when compared to the patient's birth chart, would establish the necessary treatment required, predict periods of potential crisis, and warn the patient of future susceptibilities to disease.

But knowing what was wrong with the patient was only half the problem. Finding the cure was the other half, and this too involved a knowledge of the sky and astrology, since the correct medicine given at the wrong time could be dangerous or even fatal.

Herbs were considered to have great medicinal properties for certain diseases if collected at the proper time. These times, in

terms of the day of the week and hour of the day, were determined
by the celestial object associated with the patient's condition.

The "art" of healing was complicated even more by the moon.
As the moon completed its 29.5-day tour through the zodiac signs,
it was supposed to influence the part of the body associated with
the individual sign it inhabited. This, of course, also influenced
the proposed medicine and time of treatment.

Astrology is no longer considered an essential part of medicine.
As the practice of medicine proceeded along scientific lines, it cast
off the pseudoscientific aspects of its early origins. But souvenirs
of former times when medicine was influenced by myth and astrology endure. The modern pharmacy is still identified by an Rx
symbol. This symbol for prescription drugs is a modification of
an Egyptian symbol representing the injured, but healing, eye of
Horus.

The sky has also had a long association with music. The perceived
harmonies of the universe long ago provided the inspiration to
seek musical harmonies in the sky like those discovered on earth.
The harmonies were sought in the mathematics of the structure
and dynamics of the solar system. The magic sky exerts a strong
influence on music today. It serves as a constant inspiration to
the lyrics of contemporary music.

Twenty-six centuries ago the Greek philosopher Pythagoras
discovered that the pitch of a note depends on the length of the
vibrating string that produces it. A harmony of sound was then
achieved by the use of numerical ratios of the vibrating string
length. A 2:1 ratio produced an octave, 3:2 a fifth, 4:3 a fourth.
It was the beginning of music and science, the discovery of a successful relationship between sensations and mind, quality and quantity.

Thinking all things could be understood ultimately by numbers,
Pythagoras looked to the sky. He found movement and numerical

relationships again. The sun, moon, and each planet were assigned
a number by Pythagoras based on their movement, and this movement caused a celestial sound. The numerical ratios of the numbers
assigned to the movement of the different celestial wanderers produced a perfect heavenly harmony that could never be achieved
by terrestrial strings, reeds, or horns.

Ordinary humans did not hear this musical gift from the
universe surrounding them. You had to be special. One of those
special people was the seventeenth-century German mathematician-
astronomer-mystic Johannes Kepler (1571–1630).

Kepler believed if you could look into the mind of God you
would see mathematics. He was therefore not surprised to look
into the universe and see numbers. Kepler saw numbers in the
celestial velocities of the planets that produced the terrestrial
harmonies found pleasing to the ear in earthly music.

Kepler's harmonies also reflected a fundamental change in
the view of the universe. Pythagoras's musical harmony was geocentric. It was based on measurements from an earth-centered universe. Kepler's harmony was heliocentric. It was based on measurements from a sun-centered universe.

A music of the planets became audible to everyone early in
the twentieth century through the work of another mystic, the
composer Gustav Holst (1874–1934). Holst's most popular work,
"The Planets," was not based on mathematical considerations of
the orbits or distances of the planets, but on themes corresponding
to the astrological associations of the planets. Holst "sound painted"
the planets to produce the first pieces of music to create the emotions
now associated with "outer space."

Sky phenomena have also provided the inspiration for musical
"text painting," providing words and visions to accompany the
melody. Think of the number of songs using sun, moon, and star
imagery that have lasted for decades, songs like "You Are My
Sunshine" (1940), "Shine On Harvest Moon" (1944), and "Star
Dust" (1929). Country and western music, because of its original

outdoor orientation, has made the most frequent use of sky imagery.

There truly is music in the night sky, but it is not found in the celestial harmonies sought by Pythagoras and Kepler. In 1977, two NASA Voyager spacecraft began a trip to the stars. Onboard, they carried an 87.5-minute recording of representative earth music, earth harmonies. Surprisingly few of the pieces contain astronomical references. The "stars" of this recording are the likes of Bach, Beethoven, Mozart, and Chuck Berry.

Someday, near some far distant star, another people may discover a celestial harmony. The Voyager record may be their first positive proof of the existence of another life form in their universe. Music and numbers will combine. And a cycle will have been completed.

The magic sky of primitive wonder and worship created by the ancient stargazers is still alive in the twentieth century. It lives in words, terms, and phrases of suspected or unmistakable astronomical origin or association.

The automobile industry has often used the sky to market its products. It has created automobile models named Polaris (the North Star), Vega (the Summer Triangle star), Nova (an exploding star), Pulsar (a neutron star), and Taurus (a constellation). The name of the Japanese automobile manufacturer Subaru is the Japanese equivalent for "Pleiades," an open star cluster in the constellation Taurus the Bull. The company's logo features the Little Dipper arrangement of the six visible stars of the Pleiades star cluster.

Places have also drawn upon the sky for their identity. There are towns in the United States named Mars (Pa.), Jupiter (Fla.), Neptune (N.J.), and Earth (Tex.), as well as Star City (Ariz.), Skyland (N.C.), and Moon (Pa.) Apparently, the only town named after a constellation is Orion, Illinois.

The radioactive elements uranium, neptunium, and plutonium were all named after planets that were discovered with the use of a telescope. The new planets, like the naked-eye planets, were named after the gods of myth. Elements 93 and 94, neptunium and plutonium, lie beyond element 92, uranium, on the periodic table just as the planets Neptune and Pluto lie farther beyond Uranus in the solar system. Also, Walt Disney's dog Pluto made his film debut suspiciously near the time of the discovery of the solar system's outermost planet of the same name in 1932.

Many of the words and phrases we use in everyday language have astronomical roots. The origins of some of the words and phrases are obvious, while the origins of others are not. And the origins of some are suspicious, or not well established. The moon has produced a number of these words and phrases.

The appearance of the full moon, for instance, has produced a word that attempts to describe an act that could be called mildly antisocial. "Mooning" is the purposeful act of briefly exposing one's posterior to an unsuspecting observer. It is usually done for its surprise and shock value out the window of a moving vehicle such as an automobile or school bus.

The appearance of the full moon has also produced a word that attempts to describe a more sociable event. It is a word whose origin is most probably rooted in a subtlety in the appearance of the June full moon, but which has been extended to include like events throughout the year. The subtlety requires some explanation.

You may think if you've seen one full moon, you've seen them all, but this is not so. The full moon rises and sets in different places each month, and it reaches a different maximum altitude during the different months of the year. Most importantly, the full moon's color changes during the different months of the year.

The movement of the full moon each month is the same as that of the sun at the opposite time of the year. This is because the sun and the full moon are in opposite parts of the sky. Since

June is the month of the summer sun, the June full moon will closely mirror the movement of the winter sun of December. The June full moon will rise far to the south of east. It will not be very high when it reaches its maximum altitude, and it will set far to the south of west.

As the light from the full moon travels through the earth's atmosphere, it interacts with the atoms and molecules that constitute the earth's atmosphere. The blue end of the white light spectrum is scattered by the atmosphere to a greater extent than the red end of the white light spectrum. This causes the rising and setting moon to take on a familiar reddish appearance when it is near the horizon.

Since the full moon in June is always low in the sky and therefore near the horizon, its color is subtly, and most importantly, continuously affected by the passage of its light through the atmosphere. The June full moon therefore takes on a golden appearance like that of honey.

The June full moon was referred to as a "honey moon" and people who married in June, the busiest month for such events, consequently had a "honeymoon." The term was then supposedly broadened to relate to the post-nuptial period for couples who married at any time of the year.

There are other explanations for the term of course. A more skeptical explanation holds that the first month of marriage is the sweetest, but that the nuptial affections thereafter "wax" and "wane" as does the moon in going through its phases.

Many English words have meanings derived from the astro-logical characteristics given to the planets and constellations. Those who are mercurial have as quick and changable a character as Mercury the messenger god. All things venereal pertain to physical sex. Venus was the goddess of love. Things that are martial have to do with war. Mars was the god of war. Those who are jovial enjoy themselves like Jupiter, the supreme god. Those who are saturine are gloomy and uncommunicative. Saturn was the slowest moving and most distant planet. Those who are capricious are as impulsive and unpredictable as Capricorn the Goat-Fish.

Many English phrases have astronomical connotations. If something happens "once in a blue moon," it happens as rarely as the occurrence of two full moons in the same month. Blue moons occur at approximately 30-month or 2.5-year intervals. If someone is in "seventh heaven" they are in the highest perceived shell of an early cosmological model of the universe. They are in the abode of God and angels. To be thought of as having one's "head in the stars" is to be thought of as an impractical dreamer. The expression "by Jiminy" is an old fashioned appeal for assistance from the Gemini twins. The expression "by Jove" is an old fashioned appeal for assistance from Jupiter (Jove). To wish someone *mazel tov* is to wish them good luck in Yiddish. The phrase "mazel tov" means "may your planetary influences be favorable."

Some of Western history can be found captured in the names of stars. The early Greek influence can be seen in the names of some of the brighter night stars such as Sirius, Capella, and the Pleiades. Greek astronomy was part of the booty collected by the Romans as a result of their rise to dominance in the Mediterranean. The Romans had interests in other areas and contributed little to the advancement of astronomy. They did give the naked-eye planets their names, however. With the near total conquest of the Mediterranean by the followers of Islam, astronomy passed to another people.

The Arabs preserved Western civilization while Europe suffered through the Dark Ages. What we know of the ancient world and astronomy is largely due to the Arabs' preservation of this information. Europe eventually recovered its lost heritage to begin the Renaissance. But the astronomy returned to Europe was not returned intact.

Arabic astronomers had little influence on astronomical theory. They mostly translated and studied the manuscripts of the people they had conquered. But Arabic astronomers had a great influence on the language of astronomy. The names they gave to many of the stars have remained. These star names are a legacy derived from the almost one thousand-year-long period of time that the Islamic civilization controlled what was known of the world.

One has only to look at our early winter night sky to appreciate the Arabic influence in star names. In the very early evening north-western sky, the three bright stars of the Summer Triangle can still be seen. These stars, Vega, Deneb, and Altair, are all de-rived from Arabic words.

The early evening southern sky is dominated by the four bright stars of the autumn asterism ("a nickname or unofficial name for a prominent group of stars") known as the Great Square of Pegasus. These four stars, Alpheratz, Scheat, Merkab, and Algenib, also have their origins in Arabic.

In the later evening sky, Aldebaran, the eye of Taurus the Bull, and the very bright winter constellation of Orion the Hunter will appear above the eastern horizon. The Orion rectangle is outlined by the stars Betelgeuse, Bellatrix, Rigel, and Saiph. The belt stars within the rectangle are named Alnitak, Alnilam, and Mintaka. We owe these star names to Arabic also.

What we call something says a good deal about that some-thing. Decisions are involved. One name is accepted, other names are rejected. "What's in a name?" asks Shakespeare. "That which we call a rose by any other name would smell as sweet." In principle, yes. In practice, no.

Names are not simply names. They have value. Things can have a "good name," "a big name," or a "brand name." And having a "bad name" can hurt almost as much as sticks and stones. Name wars have been fought across the heavens.

During the early part of the seventeenth century an effort was made to remove all the pagan influences from the sky and to "Christianize" the constellations. Julius Schillerius started the movement in 1625 with his "Coelum Stellatum Christianum." He was aided by Jocabus Bartschius, who produced celestial globes bearing Christian constellations. Unfortunately for the movement, Schillerius and Bartschius did not always agree as to what name should be given to particular constellations. The situation was made even worse by Wilhelm Schickard, who introduced another set of Christian names at the same time. Divided they failed.

If they had won, the twelve zodiac constellations would have been replaced by the twelve apostles. The then-known planets would have become Elias (Mercury), St. John the Baptist (Venus), Joshua (Mars), Moses (Jupiter), and Adam (Saturn). The northern circumpolar sky would have been dominated by the bright stars of St. Peter's Fishingboat (Ursa Major—the Big Dipper) and St. Mary Magdalen (Cassiopeia). The fainter circumpolar constellations would consist of St. Michael (Ursa Minor—the Little Dipper), St. Stephen (Cepheus), and the Innocents (Draco).

This was not the only attempt in history to change the names of the stars and constellations. In 1807, the University of Leipzig formally declared that the stars that form Orion's belt and sword would henceforth be known as the constellation "Napoleon." They were probably inspired by the fact that Napoleon's army had recently taken control of Leipzig.

Napoleon's enemy, the British, characteristically rose to the occasion. They proposed that the entire Orion constellation be renamed "Nelson" in honor of the British naval hero who died from wounds received during the 1805 Battle of Trafalgar, in which the French had been defeated.

Orion remained Orion. The belt and sword stars remained the belt and sword stars.

When William Herschel (1738–1822) made the first discovery of a planet with the telescope in 1781, he choose to name the planet "Georgium Sidus," after George III, the king of England. There were objections, of course, since George III was not a star and George III had nothing to do with the planet's discovery.

"Herschel" was a second choice, but it was eventually decided that planets should be named after ancient gods and goddesses. "Georgium Sidus" became "Uranus" in honor of the ancient Greek god of the sky. But this was not quite the end of it.

In 1846, another planet was discovered, but there was a controversy as to who deserved credit for the discovery. The French suggested the new planet be called "Leverrier" after the French astronomer involved in the controversy, and that Uranus be renamed "Herschel." It was decided by all concerned that "Neptune" was the better choice for the "new" planet. Neptune was the ancient Roman god of the sea.

The moon has served as a particularly interesting nomenclature battlefield. In the middle of the seventeenth century, Johannes Riccioli, an Italian Jesuit priest, named the craters of the moon after famous philosophers and scientists. Names were assigned chronologically from north to south. But Riccioli made a few value judgements in his assignments. The nonchurch-favored Galileo was assigned an insignificant crater in the Ocean of Storms, whereas the church-favored Tycho Brahe was assigned the brightest crater on the moon.

The space age created a unique set of nomenclature problems. It also made the moon a battleground once again. There were no accepted rules for naming features discovered by spacecraft. The Russians got the jump on naming features on the far side of the moon by virtue of the photographs from Luna 3 in 1959. The United States caught up with the Russians in the "name race" by virtue of the Apollo lunar program.

Space age discoveries made by spacecraft flybys come in large numbers over short periods of time. Problems arise from the need to quickly name objects and features for reference at the time of a flyby. It is not always possible in a practical sense to wait for an official sanctioning or discussion of a name.

Certain traditions have been established for the naming of celestial objects within the solar system. Comets are given the names of their discoverer. Asteroids are named by their discoverer, who can attach any name desired to the asteroid. The official names of the craters and formations on the planets and their satellites are ultimately determined by the International Astronomical Union, but the IAU has followed the tradition used to name features of the moon by honoring outstanding humans from all endeavors.

As long as you are not concerned with what other people do, the bottom line in nomenclature is that anyone can name anything any name they want. And it has been discovered that sometimes people will even pay for this free privilege. A controversy has arisen challenging the ethics of those who have taken advantage of this situation.

"*Caveat emptor!* Let the buyer beware!" This is an ancient Latin admonition, one of the earliest consumer advocacy slogans. But every buyer does not heed this advice. That is why the advice has outlived the language.

What happens when the advice is not heeded? An English proverb explains the individual result: "A fool and his money are soon parted." Phineas Taylor (P. T.) Barnum summarized the extent of the market when he proclaimed, "There is a sucker [read fool] born every minute."

Many commercial successes have been based upon these two marketing principles. The beneficiaries of such successes have often been highly criticized as unethical. They often respond by pointing out that they are involved in completely legal operations. No laws are broken, and there is no harm done if no one is hurt.

The criticized business is that of selling star names. The star naming controversy involves the question of who has the right to name stars. In the past the answer was easy. Astronomers named stars. The stars and celestial objects of the night sky were part of their protected territory. The stars were the astronomers' turf.

Most of the bright stars of the night sky were named by the unknown stargazers of ancient times. It was probably not done by committee consensus. The proper names of the bright stars were most likely fixed by the prestige of the namer, or from their appearance on a popular star chart. The former was the case, in more modern times, with Regulus in Leo the Lion. The star was given its name by the sixteenth-century Polish astronomer Nicholas Copernicus.

The latter was the case with star names that mix the letters of the Greek alphabet with the Latin possessive of the constellation within which the star is found. Johann Bayer (1572–1625) of Augsburg, Germany, published a popular celestial atlas of sixty constellations in the seventeenth century. He called the brightest star in any constellation "alpha," the second brightest star "beta," the third "gamma," and so on through the twenty-four letters of the Greek alphabet. After the Greek alphabet was exhausted, the letters of the Roman alphabet were used.

The Latinized version of the encompassing constellation was added to distinguish between the alphas of different constellations. Thus, Alpha Leonis is Regulus, Alpha Virginis is Spica, and Alpha Bootis is Arcturus.

Every star in the sky has actually been named in the sense that every star visible can be identified by a series of numbers that locates its position on a grid that covers the entire sky.

It is interesting that in only one case have stars been named after a human being. The person responsible for this anomaly was Nicolaus Venator, a "no-name" assistant draftsman who was preparing a star chart for a "big-name" astronomer. Venator quietly attached the names "Sualocin" and "Rotanev" to two stars in the small constellation Delphinus the Dolphin. It took many years before someone recognized the origin of the star names. Sualocin is Nicolaus spelled backwards, and Rotanev is Venator spelled backwards.

Stars have been sold to the public in the past, and even the craters of Mercury and the moon have been sold to the public in the past, but only as fund-raising efforts to benefit a nonprofit organization or public planetarium. No one has complained about these activities even though the buyers have not acquired legal ownership.

The for-profit, star-naming businesses sell a personalized plaque identifying the star and its proposed name. They also include a small star chart. This is as innocent as the adoption papers for a Cabbage Patch doll. But the name is also entered in a book that is filed with the Library of Congress or placed in a vault in Switzerland. Here is where the possibility of deceit exists.

Caveat emptor to anyone who thinks these books have any official astronomical standing. Caveat emptor to anyone who thinks these books will ever be opened by an astronomer, or that the enclosed star names will ever be used by an astronomer, or anyone else other than the one so honored. It is very difficult to change long-established traditions.

Star groups can, however, be known by different names at different times and to different cultures. But these names are more appropriately referred to as asterisms or nicknames. The most famous case is that of the seven brightest stars of the constellation Ursa Major—the Large Bear. In the United States this star arrangement is known as the Big Dipper. In England, it is known as the Plough. The same seven bright stars have also been identified with various vehicles of transportation over periods of time. In early England, they were known as King Arthur's Chariot or Wain; in Ireland as King David's Chariot; and in France as the Great Chariot. The Danes, Swedes, and Icelanders once knew the stars as Thor's Wagon.

Another remnant of the magic sky of primitive wonder and worship can be found in the symbols used on the flags of many nations. The American flag holds the record for stars with fifty, while Brazil is a distant second with twenty-three stars. Dominica is third with ten stars, while Venezuela is fourth with seven stars, one more than the host of countries using six stars.

The most widely depicted constellation on world flags is found in the night skies of the Southern Hemisphere. The bright stars of Crux, the Southern Cross, appear on the flags of Australia, New Zealand, Western Samoa, and Papua New Guinea.

After stars, the next most frequent astronomical object appearing on world flags is the crescent moon. It usually appears with a star on the flags of Muslim countries. The crescent moon and star serve as a reminder that the sky is a calendrical guide to fix the dates for the observance of important Muslim religious duties, such as the time for the pilgrimage to Mecca and the fasting month of Ramadan.

The sun is suprisingly the least favored celestial object to be depicted on a world flag. It only appears on the flags of Argentina, Japan, the Philippines, and Uruguay.

What is true of the flags of the world is basically true of the flags of the fifty states of the United States. Almost half the state flags feature stars, the sun, or the moon. The only constellation is found on the flag of Alaska, which has a representation of the seven Big Dipper stars in Ursa Major along with Polaris, the North Star.

Philatelists (stamp collectors) have also been exposed to the lasting influence of the magic sky. A worldwide interest in the naked-eye sky, astronomical events, and the twentieth-century "wonder and worship" of space exploration can be seen in the more than 130 countries that have issued stamps with astronomical and space exploration themes. Stamps have been issued depicting constellations and celestial objects as well as to commemorate the occurrence of solar eclipses and specific achievements in the realm of space technology. The full breadth of astronomical stamps includes representations of famous astronomers, telescopes, observatories, and archaeoastronomical sites.

The night sky has not strongly influenced the world of art, with the obvious exception of space art and science fiction illustration.

PRECESSION OF THE EQUINOX

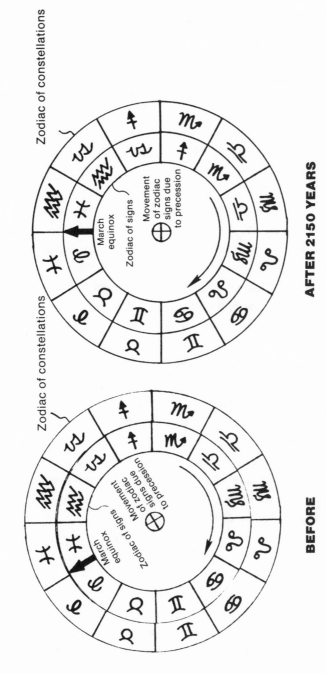

BEFORE

AFTER 2150 YEARS

The zodiacs of signs and constellations.

One of the problems involved with painting the night sky involves the inability of the artist to see what he or she is doing while in the dark. Another problem involves the representation of the night sky. A white dot on a black background does not, for some reason, completely convey the aesthetic nature of our celestial canopy.

There was one famous nineteenth-century Dutch painter who managed to overcome these difficulties. He set up his easel on the pavement outside of the Cafe de l'Alazar in Arles, France, and illuminated his canvas with candles attached to his hat. His name was Vincent van Gogh. Since at this period of his life he belonged to the Impressionist school of painting, he had no problems in transforming the stars into bright blotches of color.

Van Gogh painted a series of "starry night" scenes. The stars were depicted as large bright, whirling entities, but at no time did van Gogh attempt to accurately represent the stars in their constellation configurations. Instead, he showed the night sky as a dynamic and colorful, spiraling scene with the magnified stars and orange crescent moon he saw in his mind. Van Gogh's paintings of day scenes were often dominated by the sun. It hangs like a bright yellow, fiery disk just above the horizon. The sun was a natural object for van Gogh to paint since yellow and gold were his favorite colors.

Unfortunately, van Gogh did not find the beauty in life he found in the sun and the night sky. The last few years of his life were spent in mental asylums. He painted his most famous "Starry Night" during one such confinement.

If van Gogh had favored the color red, he undoubtedly would have included the red planet Mars in one of his starry night scenes. He didn't, but that didn't keep Mars from affecting the arts. No planet has influenced literature more than Mars. No planet has contributed more to twentieth-century mythology than Mars .

Mars is a deserted wind-swept sphere of pink skies, empty red plains, tall mountains, deep expansive canyons, extinct volcanoes, and carbon dioxide polar caps. At the same time, Mars is the planet of the "Martians." It is the stage upon which alien civilizations compete for meager resources and survival. It is the planet whose beings sometimes look earthward with covetous eyes. Mars is a planet of literary ghosts.

Mars is in our night sky. It is seen as a bright reddish star moving through the stars of the zodiac. It is the fastest moving and brightest of the outer planets. For these reasons and its color it attracted and held the attention of stargazers.

It has a fascination today because Mars lives in our subconscious mind. It is the first place we think of when we think of life beyond earth. No other object of the night sky, with the exception of the moon, has influenced our imaginations more than the red planet Mars.

If you think of Mars, you think of Martians. This is the result of science fiction writers. Some intentionally wrote fiction, some unknowingly wrote fiction.

Vikings 1 and 2 landed on Mars in the summer of 1976. They took pictures and tested the landing site for signs of life. The planet proved to be as lonely as scientists expected. No one was home except for the ghosts. In 1982, the last of the Viking landers joined the ghosts of Mars. It no longer responded to the calls from NASA.

American astronomer Percival Lowell (1855–1916) created the first ghosts of Mars. They were the canal builders of Mars. It was over one hundred years ago that the Italian astronomer Giovanni Schiaparelli (1835–1910) reported observing an interlocking network of long, dark lines on Mars. This was 1877, the year of a favorable opposition with Mars. The red planet was near the perihelion point of its orbit at the time of opposition with the sun. It was only thirty-five million miles from earth. (During least favorable oppositions, when Mars is near the aphelion point of its orbit, it can be up to sixty-three million miles from earth.)

Schiaparelli used the word *canali,* an Italian word for "channels" meaning narrow, straight lines, to describe his observations. The word was unfortunately translated into English as "canals," implying waterways of intelligent design. And that was the beginning of life on Mars. Interestingly enough, no one seems to have tried to correct the translation. Or if they did, it was too late, once again proving that the pen is mightier than the eraser.

The existence of the canals was questioned by some astronomers and confirmed by others. Percival Lowell was one astronomer who was convinced that the canals of Mars were real and not the result of telescope quality, observing conditions, or the overactive imagination of the observer. Lowell was profoundly influenced by Schiaparelli's observations, even to the point of using his considerable wealth to establish an observatory on Mars Hill in Flagstaff, Arizona, in 1894 for the purpose of observing Mars under clear, steady, light-free skies. Lowell interpreted the canals of Mars as a planet-wide effort by the Martians to irrigate a planet dying of thirst. His books were very popular and very convincing.

Percival Lowell was also very wrong. There are no canals on Mars. Then what had the astronomers seen? No one has been able to offer a satisfying physiological or psychological explanation for the canal observations. Lowell was successful in one respect. His scientific writing succeeded in creating the Martians of science fiction. And what Lowell joined together, no amount of science will ever put asunder.

Science fiction did not begin with Lowell's Martians, but he provided the catalyst for its popularity. If science fiction is defined as a form of fiction that incorporates contemporary scientific plausibility, as opposed to pure fantasy, then Johannes Kepler can rightfully be considered the first science fiction writer.

The moon, the earth's closest neighbor in space, was understandably the first target of science fiction writers. Kepler's *Sominium* (1630) was a dream voyage to the moon. Cyrano de Bergerac used gunpowder rockets for *A Voyage to the Moon* (1650). Jules Verne used a cannon for his fictional voyage *From the Earth to the Moon* (1865).

Edgar Rice Burroughs brought adventure and excitement to science fiction with his tales set on Mars. He gave the Martians names and forms. He created the threatening world of Barsoom and populated it with violent red and green creatures with six and eight limbs. He also created the beautiful (by earth standards) Dejah Thoris, Princess of Helium, a kingdom on the red planet Barsoom, and John Carter, space adventurer from Virginia.

Orson Welles brought the Martians to earth on Halloween night, 1938, with a radio adaptation of an H. G. Wells novel, *The War of the Worlds*. The radio program, in the form of a realistic news broadcast, described an invasion from Mars of mechanical monsters. The population of Grovers Mill, New Jersey, the sight of the "landing," and the surrounding countryside went into a monumental panic as they took to the highways in search of places of safety.

Ray Bradbury brought the Martians and science fiction to literature classes with his classic *Martian Chronicles*.

The visual media completed the picture of the Martians with art and film. Science fiction artists illustrated the covers of pulp magazines with the fantastic creatures that lived within its pages. Science fiction films like *Angry Red Planet* brought life to those fantasies.

And then, of course, there was the television series "My Favorite Martian" and a Hollywood film, entitled: *Santa Claus Conquers the Martians*. What they did to the Martians is libelous.

The truth about the Martians is now recognized. *We* are the Martians. It is the destiny of earthlings to become Martians in the twenty-first century.

Some of the magic sky wonder and worship will follow. It

is not difficult to imagine a future Martian looking out at the magic night sky and without thinking recite:

> Starlight, star bright,
> First star I see tonight,
> I wish I may, I wish I might,
> Have the wish I wish tonight.

4

Clock and Calendar Astronomy

Astronomy's most important contribution to civilization, and its most important form of utilization, is time. Astronomy is all about time and the measurement of time—time past, time present, and time future. Our clocks and calendars are not just simple devices for crossing off the days, weeks, months, and years. They are measures of the changing conditions of the universe. Clocks and calendars save the phenomena. The earliest civilizations of the Eastern and Western hemispheres recognized the need to record the passage of time. It was this need that initially led to the desire to observe in detail and record the motions of the sky.

The sky was a rich source of natural cycles with which to measure time. Someone was required to keep track of the cycles because none of the cycles fit together perfectly. This someone became known as an astronomer, and what he learned of the night sky while tending his calendrical duties became astronomy. The development of an explanation of the astronomer's observations in the form of a conceptual model became the history of astronomy.

Without time there would be no civilization. There would be no condition of human society distinguished by an advanced stage of development. There would be no arts and sciences, nor would there be social, political, and cultural organization.

It has been suggested that you really do not truly understand something until you are able to measure or apply numbers to it. The opposite seems to be true of time. We have been measuring time longer than any other thing set to numbers, but we still do not fully understand it. It is impossible to define without referring to its means of measure.

Any event that occurs repeatedly at fixed intervals can be used as a measure of time. The interlude between two such consecutive events can be used as a unit of time. The most fundamental unit of time is provided by the rotation and revolution of the earth with the resultant appearance and disappearance of the sun. The first measure of time began with the sun rising above the eastern horizon. This was the beginning of "day" time. When the sun set below the western horizon, daytime ended and "night" time began. Another rising of the sun began another day. The "uncivilized" creatures with whom we share this planet also measure time in this way to regulate their activities.

The world's oldest astronomical instrument and maybe even the world's oldest scientific instrument is a simple stick rising vertically out of the ground. In scientific terms it is called a "gnomon." The word means "one who knows." A shadow is created by the gnomon at sunrise. As the sun moves, the gnomon's shadow moves. Our day proceeds as the sun moves east to south to west across the sky from its rising position to its setting position. The shadow moves from west to north to east across the ground. Today, our clock hands move "clockwise" in replication of the movement of the sundial shadow.

As the sun gets higher in the sky, the gnomon's shadow gets shorter. Local apparent noon takes place each day when the center of the sun is due south, having crossed the observer's southern meridian, and the gnomon shadow is its shortest length for the day. This is the time of the sun's crossing of the local meridian. The meridian is an imaginary line in the sky that passes from the due north point on the horizon, through the overhead zenith point, and down to the due south point on the horizon.

The meridian divides the sky into an eastern and western half.

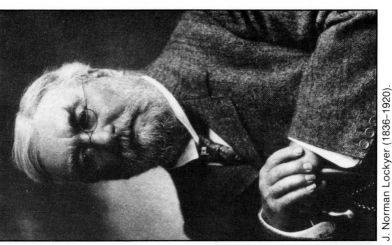

J. Norman Lockyer (1836–1920).

The Dawn of Astronomy, by J. Norman Lockyer, written and published in 1894, has provided an excellent conceptual framework within which to look at the cultural influence of astronomy. (See pp. 22-29.)

The Great Pyramid of Khufu, constructed in 2680 B.C., is the largest pyramid ever built. (See p. 126.)

The four 65-foot-tall seated figures of Ramses II fronting the temple at Abu Simbel in the Upper Nile Valley. (See p. 127.)

One of the central buildings in Chichen-Itzá, an abandoned Mayan city in the Yucatan jungle of Mexico. This 78-foot-tall temple-pyramid is dedicated to Kukulcan, the Feathered Serpent. (See p. 129.)

El Caracol ("the snail"), another structure at Chichen Itzá, was so named by Spanish conquerors because of its unusual cylindrical tower. (See p. 130.)

Stonehenge III, as seen from the vantage-point of the Heel Stone. (See pp. 119–124.)

The Heel Stone of Stonehenge, set in position to act as a marker for the rising summer solstice sun as seen from the center of the Stonehenge circle. (See p. 120.)

The Samrat Yantra ("supreme instrument") in Delhi, one of a series of large sundials built by the early eighteenth-century Mogul emperor, the Maharajah Jai Singh. (See p. 182.)

Local apparent noon measures the time of the sun's passage from before the local meridian to after the local meridian. It is from the sun's crossing of the meridian that we derive the familiar A.M. and P.M. divisions of the day. The term A.M. means "ante meridiem" or before the meridian, while the term P.M. means "post meridiem" or after the meridian. Since the interlude between two consecutive sunrises or sunsets changes over the year, the meridian transit provided a more even measure of the day upon which to impose artificial divisions of time.

The first improvement to the gnomon was undoubtedly the addition of a calibrated plate to mark off intervals in the movement of the gnomon's shadow. The combination produced the sundial clock and a means of organizing and regulating the affairs of mankind. The Sumerians are credited with dividing day and night into twelve hour units. Twelve was considered a special or lucky number. In the beginning, all days and all nights were each twelve hours long. An hour of daytime in the summer was originally much longer than an hour of daytime in the winter.

The length of the noon shadow of a gnomon changes over the course of a year. The noon shadow is shortest at the beginning of summer on the June solstice, because the sun is at its maximum yearly noon altitude on this day. The noon shadow is longest at the beginning of winter, on the December solstice, because the sun is at its minimum yearly noon altitude on this day. The sundial served as a daily clock and yearly calendar, two essential measures without which civilization could hardly have advanced beyond a most primitive level.

The measure of day units produced a problem. They accumulated too quickly. Another cycle, a longer cycle, was needed. It was found in the waxing and waning of the moon. The "month" is based on a "moonth," the time interval of 29.5 days between the recurrence of the same phase of the moon. The earliest known calendars were lunar calendars. Islam and Judaism still reckon their religious year by lunar calendars.

The Islamic calendar is strictly a lunar calendar. It consists of twelve months that alternate between 29 and 30 days so that the average month becomes 29.5 days long. The years are numbered from July 16, 622, the date on which the prophet Mohammed fled from Mecca to Medina in order to escape from being assassinated. The A.H. designation attached to the Islamic year number stands for "Ab Hejira" or "from the flight."

The Islamic year varies between 354 and 355 days as compared to the seasonal year of 365.25 days. Since the Islamic year is shorter than the seasonal year, the Islamic New Year and the fasting month of Ramadan will move backwards through the seasons ten or eleven days each year. It will move completely through the seasons every 32.5 years.

Astronomy is closely tied to many Islamic religious practices. For instance, the important month of Ramadan does not begin until sunrise the morning after the first thin waxing crescent moon, following the new moon, is seen in the sky by two different observers. Astronomers at Arab observatories have carefully studied the difficult problem of predicting the first appearance of the thin crescent moon because of its importance to the Islamic calendar. Their predictions are respected, but they are always confirmed by observation. Other Islamic uses of astronomy include the fixing of prayer times and the determination of the direction of the holy city of Mecca for the saying of prayers.

The Jewish calendar is an older calendrical system that also goes back to the days when the moon was used as the primary unit of time. The first day of the Jewish month begins at sunset on the day of the new moon. The day begins at sunset because the Book of Genesis states "the evening and the morning were the first day."

The Jewish calendar is not a strict lunar calendar since it is also tied to the seasons. The calendar year count begins from a calculation of the date of creation, this event having been established as having occurred in the fall of 3761 B.C. according to the present Gregorian calendar. The Jewish calendar also is comprised of alternating months of 29 and 30 days to produce an average month of 29.5 days equal to the lunar cycle of phases. This also produces a calendrical system ten or eleven days short of the seasonal cycle, but in this calendrical system a correction is made by the inclusion of seven thirteen-month years within a nineteen-year period.

The ancient Chinese lunar calendar was also correlated to the seasonal year by the addition of extra months when needed. This calendar reform dates back to 2357 B.C. when the Emperor Yao became dissatisfied with the traditional lunar calendar and ordered his astronomers to devise a calendar system that followed the seasons.

The last reform of the Chinese lunar calendar took place in 104 B.C. New Year's Day by this reform was fixed to always begin on the day of the first new moon after the sun enters the astrological sign of Aquarius. This fixed the date of New Year's Day as no earlier than January 21 and no later than February 20. It also tied the Chinese calendar to astrology. Today the sun is actually in the constellation Capricornus during this period of time, but two thousand years ago when the calendar reform was instituted, the sign of Aquarius and the constellation Aquarius were aligned. The misalignment between the astrological signs and constellations is due to a slow conical motion of the earth's axis of rotation called precession.

Native American Indians of the northern and eastern United States used full moons as a calendar. The name of each of the full moons was associated with practical activities and climatic conditions, and each month was known by the full moon that occurred within it.

The full moon that occurred at the time when deer were fattest was the "hunter's moon." This full moon was followed by the "beaver moon," a time to set traps before winter set in. The winter months of December, January, and February were marked by the "cold moon," the "wolf moon," and the "snow moon." Spring was a busy season and this was reflected in full moons of spring having several names. The March full moon was the "crow moon" or "sap moon." April brought the "sprouting grass moon" and May the "flower moon" or "corn planting moon." June was the "strawberry moon," July was the "buck moon," and August was the "green corn moon." Which brings us to the most famous of all full moons, the "harvest moon."

The harvest moon is the full moon that occurs nearest to the day of the September equinox, the first day of autumn. This full moon received its name from a phenomenon related to the moon delay, which is a measure of the difference in the time of moonrise from night to night. The average moon delay at latitude 40 degrees north is 54 minutes. This is the yearly average of the extremes. There is a time of maximum delay and a time of minimum delay.

During the winter and spring seasons the moon delay is longer than average, and during the summer and autumn it is shorter than average. The moon delay is seventy-five minutes at its longest in January and twenty-two minutes at its shortest in September. A minimum moon delay meant that a bright, shadow-casting, nearly full moon was in the sky for several evenings. This extra light was used to great advantage by farmers to extend their fall harvesting hours. So the full moon nearest the September equinox became known as the "harvest moon."

The use of the harvest moon has disappeared with the advent of tractor headlights, but its memory lives on in song. "Shine On Harvest Moon" is a golden oldie written by Nora Bates (nee Dora Goldberg), who also gave us another autumn classic, "Take Me Out To The Ballgame." The harvest moon song first appeared in the Ziegfeld Follies of 1908, and later in Nora Bates's 1944 film autobiography, *Shine On Harvest Moon.*

The moon has also influenced the dates of moveable religious celebrations. The Christian celebration of Easter, for instance, can occur as early as March 22 or as late as April 25 because the date of Easter is tied to a phase of the moon. Easter is celebrated in Catholic and Protestant churches on the Sunday following the first full moon that occurs after the day of the March equinox. If the first full moon occurs on a Sunday, then Easter is celebrated the following Sunday.

The question of when Easter should be celebrated arose in the early days of Christianity. The question was resolved by the first ecumenical synod called in A.D. 325 by the Roman Emperor Constantine. The Council of Nicea, in trying to set the Easter date in keeping with the biblical sequence of events, produced the above formula with the added stipulation that Easter should also be celebrated after the Jewish Passover.

Passover is a moveable calendar date that begins on the four-teenth day of the first month of the Jewish lunar-solar calendar. Never trust a committee to arrive at a simple solution to a simple problem. The Roman Catholic and Protestant churches later abandoned the Jewish Passover rule. The Eastern Orthodox church did not, and that is why all Christian churches do not always celebrate Easter on the same Sunday.

The full moon was involved in the formula for practical reasons. It provided the pilgrims who traveled to Jerusalem with a bright night sky in order to ensure their safety.

The biggest disadvantage of a lunar calendar is that it is not tied to the seasons. It was therefore inevitable that a solar calendar would be developed. The seasonal year was first determined to be nearly 365.25 days in number by the Egyptians, who marked the new year with the date of the helical rising of the star Sothis. This is the star known today as Sirius, the brightest star in the night sky. A helical rising refers to the date when a celestial object first becomes visible above the eastern horizon before sunrise. On days prior to the helical rising, the object is lost in the glare of the sun.

Julius Caesar introduced the solar calendar to Western civilization in 46 B.C. He used a consultant named Sosigenes, who was an astronomer from Alexandria. It was he who suggested that the Roman Empire adopt the three thousand-year-old Egyptian calendar that was based entirely on the motion of the sun. In order to get the new Julian calendar started properly, it was necessary to do some funny things with the year 46 B.C. The result was that this year consisted of fifteen months and 445 days. Once the "Year of Confusion" was over, everyone sat back and relaxed. They now had a good calendar.

The years of the Julian calendar were not perfect, however. They were 11 minutes and 14 seconds too short. This seems rather insignificant, but because of this, by the year 1582, the first day of spring had moved forward from its intended date of March 21 to March 11.

In 1582, Pope Gregory XIII did what Caesar did before him. He went to the astronomers to revise the calendar. Following the astronomers' advice, Pope Gregory brought the first day of spring back to March 21 by dropping ten days from the 1582 calendar year. October 4, 1582, was followed by October 15, 1582. The leap year of 366 days and its rules of application was another reform introduced to the calendar at this time.

The Gregorian calendar is not perfect. It is in error by one day every 3,300 years. This is something to worry about later.

The Gregorian calendar was adopted immediately in all Roman Catholic countries. Quite predictably, Queen Elizabeth I did not follow suit. The Julian calendar continued to be used in Great Britian and her American colonies until 1752. In that year the Gregorian calendar was finally adopted. The switch from the one calendar system to the other was made by the elimination of the eleven days between September 2 and September 14. In 1752, thirty days hath not September.

The switch was not without its problems. It was greeted by riots and angry mobs yelling, "Give us back our eleven days." The English and their subjects were a jolly bit annoyed at having to pay monthly bills with only three weeks' wages. Eventually things quieted down and everyone accepted the calendar system. They accepted it so well they even changed the dates of some past events to conform with the Gregorian calendar date. George Washington was born on February 22, 1732, by the Gregorian calendar, which was referred to as the New Style calendar date, but he was born on February 11, 1731, by the Julian calendar, which was referred to as the Old Style calendar date.

The eleven-day difference in Washington's birthday in the two calendar systems looks more like a year because of a parallel change in the celebration date for New Year's Day. The rise of Christianity had its effect upon the celebration date for the beginning of the year. A successful attempt had been made to substitute a Christian celebration date for the pagan new year date of January 1. Christmas, Easter, and Annunciation Day had been strong candidates, but Annunciation Day had won. At the time of the calendar reform, the new year began on March 25, Annunciation Day. Part of the 1584 calendar reform involved a decision as to which date would constitute the beginning of the new year. Pope Gregory was pressured to return to the January 1 date. He did.

The month of January was chosen originally to begin the year because of its association with Janus, the mythical Roman god of doors. Janus was the two-faced god looking forward and backward who presided over entrances and exits, beginnings and endings. January was a good choice for the first month of the new year. The beginning of a new year was a time to look forward with anticipation at what was to come, and to look back in memory at what had been.

The year to begin a calendar count is important. In A.D. 533, a Roman monk named Dionysius Exigus petitioned to have the dates in history recorded in the Julian calendar system from the time of the birth of Jesus. He suggested B.C. to signify dates "before Christ" and A.D. to signify later dates as *Anno Domini,* or "in the year of our Lord."

Exigus used the information that Jesus was born in the twenty-eighth year of the reign of Caesar Augustus to fix the beginning year of the new system. However, unknown to Exigus, Augustus had ruled for four years under the name of Octavian. The name Augustus had been conferred upon him as an honor by the Roman Senate. So, Jesus was probably born in the year now identified as 4 B.C. instead of A.D. 1. (There is no zero year.)

The ultimate starting date for a year count system would be the year of the creation. This occurred on Saturday night, October 22, 4004 B.C., according to the deliberations of James Ussher, the seventeenth-century Archbishop of Armagh, Ireland. Humanity was created at 9:00 A.M. on the following Friday, and the great flood occurred 1,556 years later. Ussher's famous age and history of the earth and universe were carefully calculated from his study of references in ancient scriptures.

The 4004 B.C. date was widely accepted and subsequently inserted in the margins of authorized versions of the Bible. It soon acquired the infallibility of the Scriptures. The acceptance of the six thousand year age for the earth and universe had an impact beyond theology, however. It placed severe constraints on geological and astronomical thinking for well over one hundred years. The geologists had the most difficult time because they were just beginning to accumulate evidence that strongly indicated that the earth was much older than six thousand years.

Because time can be measured by the interval between any two consecutive events, there can be, and have been, many time measures. Some have survived and some have not. The surviving time measures proved useful for diverse cultures through the ages. The sun, and the moon to a lesser degree, have proven to be survivors. The nonsurviving time measures had a specific usefulness to a specific locality or epoch. They lasted only as long as the culture or epoch that use them.

During the days of the Old Kingdom pharaohs of the third millennium B.C., the Egyptian stargazers anxiously watched the early morning dawn sky for the day of the helical rising of Sirius. Prior to this day, Sirius was lost in the glare of the rising sun. But as the earth revolved counterclockwise around the sun, the sun slowly appeared to change its position in the same counterclockwise direction with respect to the background stars. Eventually, the sun moved far enough away from Sirius for it to appear in the eastern dawn sky just before the morning sun and thus achieve a helical rising.

The Egyptians were concerned with the helical rising of Sirius because it coincided with the time that the Nile River overflowed its banks to disperse new agricultural life to the land. The new life came in the form of a rich silt. The Greek historian Herodotus defined Egypt as "the gift of the Nile." All of life in ancient Egypt depended upon this yearly flood of the Nile.

Sirius was known in the days of the Old Kingdom as the "Nile Star." The helical rising of Sirius announced the eminent dispersal of that yearly gift. It is of little wonder that the Egyptian administrative bureaucracy used the helical rising of Sirius as the beginning of their calendar year.

But the Sirius calendar was flawed. It was doomed by time. The Sirius–Nile River connection was a temporary coincidence. It no longer exists today. The helical rising of Sirius has moved from the time of the summer solstice to the middle of August due to the precessional motion of the earth's axis. It no longer predicts the flooding of the Nile. Another change has also taken place. The Nile River no longer overflows its banks. The Aswan High Dam now controls the distribution of the life-giving water of the world's longest river.

Perhaps the most complex calendar system ever developed was that of the Mayans, a civilization that prospered in the Yucatan peninsula of Mexico before A.D. 1000. Most of what is known about Mayan interest in astronomy comes from books called codices. In 1562, the Spanish conquistadors attempted to destroy all recorded knowledge of the Mayans' pagan past. Twenty-seven codices and the knowledge within them were burned. Fortunately, three codices survived, and they have revealed some of the Mayans' accomplishments in astronomy.

These codices reveal that the Mayans used three calendar systems simultaneously, none of which were based entirely upon the exact length of the seasonal year or lunar month. The Sacred Count was a ritual 260-day counting system based upon combinations of thirteen numbers and twenty day names. The Long Count calendar was an accumulative day count based upon the Mayan date marking the creation of the present universe.

The most unusual of the three calendars was the Venus calendar system. It was based upon the movement of the planet Venus with respect to the sun. The accuracy of the Mayan Venus calendar was not matched in Europe until several centuries later. The Mayan Venus calendar was based on the knowledge that Venus is visible in the western sky as an evening star for 250 days, and is visible in the east as a morning star for 236 days. The calendar also incorporated the knowledge that Venus is lost in the glare of the sun for ninety days when it moves from its morning star to evening star appearance, and is lost in the glare of the sun when it moves from its evening star to morning star appearance for eight days.

There have been continued attempts to change our present calendar, including a patriotic attempt by the French following the revolution of 1789. The French attempted to convert to a decimal system based on the number ten. The twenty-four hours of the day were changed to ten and the seven days of the week were replaced by a "decade" of ten renamed days. Twelve renamed months of exactly thirty days each, and exactly three weeks each, left five festival days at the end of the year. Leap years added an extra day to the year and became Olympic Years.

Year I of the Era of the Republic calendar began on September 22, 1792. The reform, however, did not catch on outside of France and consequently the experiment was dropped. The French Revolution was able to successfully introduce "liberty, equality, and fraternity" into the concept of government, but not into our calendar system. Attempts to change the established traditions in the way we cross off the days, weeks, months, and years are more likely to produce cries of "Off with their heads!"

The World calendar represents a contemporary movement to change the way we cross off the days and months. This calendar would provide us with a year of four equal quarters of three months each. The months in each quarter would be thirty-one, thirty, and thirty days long, giving us exactly ninety-one days and thirteen weeks in each quarter. Each quarter would begin on Sunday and end on Saturday. The 365th day of the year would be Worldsday, an international holiday following December 30. Leap year would give us an additional Worldsday following June 30. The World calendar is a logical way to organize time that has received a hearing by the United Nations Economic and Social Council. Off with their heads!

There have been successful attempts to change the way we count off the hours of the day. Anyone who compares the time from a sundial with the time from a wristwatch will discover that the times are not the same. There is a perfectly good explanation for this. Sundials and wristwatches do not measure the same kind of time.

Sundials record local apparent solar time. This ancient system of time measurement is based upon the observed movement of the real sun across the sky.

Local apparent noon takes place each day when the center of the sun is in the due south position and the sundial shadow is its shortest length for the day. This is the time of the sun's crossing of the local meridian.

The day-to-day crossing of the meridian by the real sun provided a satisfactory local measure of time for earlier civilizations. Travel from place to place was slow and precise measurements of time were not necessary. But time based upon the real sun does not meet the demands of the highly mobile and technical world we live in today. This is because the time interval between meridian crossings of the real sun is not constant. It varies throughout the year.

The sun's passage from one meridian crossing to the next is the result of the combined effect of the rotation and revolution of the earth. If the earth only rotated, the apparent solar day would be of constant length. It would take the same amount of time to move the sun from one local apparent noon to the next. But the earth does not just rotate, it also revolves around the sun.

The revolution of the earth around the sun is the cause of the discrepancy between sundial time and wristwatch time. This is because the earth revolves around the sun at a velocity that changes during the year. In the winter, the earth moves faster in its orbit than it does in the summer. This change in velocity causes the time intervals between meridian passages of the sun to change.

The change in the time interval of each day is impossible to duplicate with a mechanical clock. A mechanical clock set to duplicate the length of the apparent solar day in winter would be running at the wrong rate in the summer and vice versa. This was a problem when accurate mechanical clocks were developed.

The problem was solved by the abandonment of the natural, but variable, time measured by a sundial, for the artificial, but constant, time measured by a mechanical clock. All days are of equal length when measured by a mechanical clock.

Today we measure the passage of time by an invention called mean solar time. The length of the mean solar day is equal to the average or mean length of the apparent solar day extended over one year, the time for one revolution of the earth around the sun. The mean sun is a fictitious sun that moves at a constant rate around the celestial equator to produce the time recorded on our wristwatches.

While the difference in the length of any one apparent solar day as compared to the length of one mean solar day is small, the accumulated difference can amount to more than sixteen minutes. The difference between apparent solar time and mean solar time is called the "equation of time." When the equation of time (apparent time minus mean time) is positive, the apparent sun crosses the meridian before local mean noon. When the equation of time is negative, the apparent sun crosses the meridian after local mean noon.

A sundial and wristwatch will both read the same time only four times each year. During the rest of the year the sundial will be "fast" or "slow."

EQUATION OF TIME

Date	Apparent-Mean Time Difference
January 1	–3 min. 14 sec.
February 1	–13 min. 34 sec.
March 1	–12 min. 38 sec.
April 1	–14 min. 12 sec.
May 1	+2 min. 50 sec.
June 1	+2 min. 27 sec.
July 1	–3 min. 31 sec.
August 1	–6 min. 17 sec.
September 1	–0 min. 15 sec.
October 1	+10 min. 01 sec.
November 1	+16 min. 21 sec.
December 1	+11 min. 16 sec.

If the declination or altitude of the sun above and below the celestial equator is introduced as another variable, a graph is produced in the shape of a figure eight. This figure eight is called an analemma.

The old sundial system of measuring time had its influence

on the mechanical system of measuring time. We still use the A.M. and P.M. division of the day, and the hands on a mechanical clock all move in the same "clockwise" direction as the sundial shadow.

There is another reason why a sundial and wristwatch will probably not agree on the time. It is because apparent solar time and mean solar time have been defined, so far, with respect to the local meridian. The sun appears to move around the earth from east to west each day. As it does, the local time, both apparent and mean, will be different at each terrestrial longitude. Noon in an eastern location will occur before noon in a western location just a short distance away.

Sundials will record these differences due to longitude within what are considered today to be rather short distances. Mean solar time should also change due to rather short differences in longitude. This was an intolerable situation in the age that introduced rapid communications and transportation.

When the age of telegraph and railroads began, just about every town had its own local time standard. It was the time displayed on the church steeple or the town hall. Time changed continuously for a traveler moving east or west. Railroad time schedules were confusing because of this. A time standard was needed that would apply over substantial distances.

In 1883, by international agreement, the earth was divided into twenty-four standard time zones of 15 degrees each. The standard time within each zone is the mean solar time at the longitude in the center of the zone. Time from one zone to the next differs by one hour since the sun appears to move across the sky 15 degrees in one hour.

Theoretically, no location within a time zone should be more than 7.5 degrees of longitude or thirty minutes of earth rotation from the center of the zone. In practice, the edges of the time zones twist and turn to accommodate geographical and political boundaries.

A fast moving traveler proceeding westward completely around the earth could *lose* up to twenty-four hours of clock time, or one day, because the earth would take more than one rotation from one noon to the next. Westbound travelers chase the sun.

A fast moving traveler proceeding eastward completely around the earth could *gain* up to twenty-four hours of clock time, or one day, because the earth would make less than one rotation from one observed noon to the next. Eastbound travelers retreat from the sun.

Imagine two travelers leaving a given location at the same time. One travels westward around the earth and the other travels eastward around the earth. They both make the trip in virtually no time at all. The one who went west would lose twenty-four hours of clock time and return a day earlier, while the one who went east would gain twenty-four hours of clock time and return a day later. To avoid this time machine effect, the 180th line of longitude has been designated the International Date Line.

Time within this zone is the same on both sides of the line, but the western side of the line is one calendar date later than the eastern side. A traveler crossing the line moving westward has to advance one calendar date. A traveler crossing the line eastward has to subtract one calendar date. The International Date Line has an irregular outline for the same geographical and political reasons that the time zone boundaries have irregular outlines.

There is a way to remember if you should subtract or add a calendar day: "If it is Sunday in Seattle, it is Monday in Manila."

Time zones are not the only changes in "natural" solar time that have been successfully instituted. In 1907, William Willett, a London builder, decided we were wasting summer daylight by sleeping through it in the morning. He suggested all clocks be moved ahead twenty minutes per month during four months of spring and summer and then back again in the fall as a method of utilizing daylight and saving energy.

This was not a new idea: It had been suggested by Benjamin Franklin in 1784. The "daylight saving time" concept was first used in Germany in 1915. It became law in England the following year. At first it was referred to as Willett Time, but later it was changed to British Summer Time.

The United States adopted a system of daylight savings time in 1917. Clocks were put one hour ahead on the last Sunday in March and returned on the last Sunday in October. The law was repealed in 1919.

The major opposition came from farmers who had trouble getting their cows to convert to the new time, kids who refused to go to bed when the sun was up, and people who opposed any sacrilegious attempt to mess around with traditional "God's time."

In February 1942, daylight savings time was reintroduced in the United States as War Time. Clocks were kept advanced for one hour until September 1945. In the following years, the use of daylight savings time was a decision that was made by consenting adults according to the prevailing standards of the community. This created confusion, with some communities adopting daylight savings time and other nearby communities not adopting it.

The Uniform Time Act of 1966 was amended in 1987 to require all states adopting daylight savings time to make the change at 2:00 A.M. standard time on the first Sunday in April, and at 2:00 A.M. daylight savings time on the last Sunday in October. Not all states use daylight savings time. Arizona, Hawaii, and part of Indiana have decided to keep the same time year-round.

There is a way to remember if you should subtract or add an hour at the changeover time: "Spring forward, fall back."

Nature in the form of the rhythms of the day and night sky provided mankind with time. And time provided mankind with the ability to organize activities. And organization led to civilization. But in the end, nature failed. Civilization discovered technology and technology demanded rhythms more uniform and precise than those found in the day and night sky.

Nature wasn't always consistent, so she was relieved of her

duties and replaced by clocks, the mechanical analogs of nature. Clocks, and the time they measure, are more responsive to the needs of civilization.

We still think of the day as one rotation of the earth. We still think of the "moonth" as one revolution of the moon. We still think of the year as one revolution of the earth around the sun. But these thoughts are part of the measures of yesterday. They are part of our dark sky legacy.

The truth is the earth is not a sufficiently accurate clock for the world of contemporary communication and technology. The rotation of the earth is slowing down. The action of lunar and solar tides has slowed the earth an estimated three minutes over the last three hundred years.

Consequently, the earth has been replaced by the cesium 133 atomic clock, with an acclaimed accuracy of one second in six thousand years, to provide an independent uniform scale of time called International Atomic Time. Leap seconds are now occasionally introduced to bring inconsistent earth rotation time into agreement with the more consistent atomic time.

The clocks of man-made time are all around us. They are on our walls, on our desks, and on our wrists. But there is still one clock that has not been replaced by technology. It is the living clock. It is the biological clock within the bodies of living organisms.

The physiology and behavior of living organisms is controlled by an internal biological clock. Many of the rhythms of this clock are synchronized to environmental cues established by celestial events, the most obvious of which is the circadian rhythm. The circadian rhythms are biological cycles like sleepiness and hunger which repeat in approximately twenty-four-hour periods.

Other biological clocks run on longer cycles. The human female menstrual cycle, for example, is "monthly."

The changing length of the day may cause seasonal changes in human behavior. Everyone, to some extent, suffers from the winter blues. Those who have it bad become lethargic, uncommunicative, unmotivated, and unable to concentrate. The main difference between winter depression and other states of depression is that extreme winter depressives have different sleeping habits. Most depressives have trouble sleeping; winter depressives begin to hibernate. They sleep for longer periods in winter than they do in summer.

It has been found that light may have therapeutic value in the treatment of extreme cases of winter depression. Some patients have shown positive signs of depression reversal when subjected to light conditions imitating the long days of summer.

A disruption of the human biological clock gives rise to the jet lag experienced by travelers crossing several time zones in a short interval of time. The mental and physical fatigue of jet lag sufferers lasts until the biological clock becomes synchronized to the day and night cycles of the new environment.

Astronauts and cosmonauts are the supreme time zone travelers. They circle the earth in ninety-minute orbits that eradicate the physical, physiological, and social measures of time, but they bring their biological clocks with them, and these biological clocks must be considered. The space traveling human biological clock would probably adapt easiest to Mars.

Mars is the most earthlike of all the planets and satellites of our solar system. For all its differences, Mars most closely mimics our seasons on earth. If we someday attempt to terraform another world, that is, change it to an earthlike environment for human habitation, Mars would be the logical choice. An acceptable planetary motion of rotation and revolution are already in place. It would just be a case of engineering the ecosystem.

Science fiction literature and serious academic proposals have supplied suggestions on how to change Mars into a more desirable piece of real estate. The essential change needed is in the atmosphere. The thin, rust-pink, dusty atmosphere of Mars would have to be replaced with a thicker atmosphere able to warm the planet by the greenhouse effect.

Humans could feel "at home" on Mars for many reasons, including the fact that the planet has familiar naked-eye celestial rhythms. Mars rotates slower than the earth, producing a Martian day just 37.5 minutes longer than an earth day. This Martian "day" is surprisingly similar in duration to the circadian rhythm "day."

Studies have shown that our "biological clocks" run at the approximate rate of twenty-four hours and fifty minutes when deprived of outside scheduling influences. It would probably be physiologically easy for human beings to adapt to a 24-hour and 37.5-minute Martian day—especially on Monday mornings.

The close correspondence between the length of earth days and Martian days is duplicated in the tilt of the rotational axes of the earth and Mars with respect to their axes of revolution. The earth's rotational axis is tilted 23.5 degrees with respect to its axis of revolution. Mars's rotational axis is tilted 24 degrees with respect to its axis of revolution.

The axis tilt of Mars is not, however, in the same direction as the earth's axis tilt. This means the apparent pole position in the Martian sky will be different from that seen from earth. The "north star" on Mars will not be Polaris, the star at the end of the handle of the Little Dipper. The Martian north star will be Deneb, the bright star that marks the tail of the constellation Cygnus the Swan.

The similarity in the tilt of the axes will cause the sunrise and sunset points on Mars to follow the same familiar extremes at a given Martian latitude as they do on earth at the equivalent latitude. The changes in the noon altitude of the sun will also be familiar. Mars even has a "land of the midnight sun" beginning at the same latitude as on the earth.

The length of the year on Mars, however, is significantly different from the length of the year on earth. Because Mars is half again as far as the earth from the sun, it moves around the sun at a slower pace. The Martian year is 687 earth days long, but because of the slower rotation rate of Mars, the Martian year is only 669 Martian days long.

The earth's orbit is nearly circular. This makes each of our seasons of approximately equal length. The Martian orbit is more elliptical and this affects the length of the individual seasons on Mars. Autumn in the northern hemisphere on Mars will be 199 earth days long, followed by a winter of 182 days, a spring of 146 days, and a summer of 160 days.

Humans aren't the only living things with internal biological clocks that respond to the day and night cycles of the environment. The nest building and mating activities of our seasonal aviary visitors are triggered by the changing seasons.

The same is true of the cycle of perennial plants. They germinate in spring and become dormant in the fall in response to temperature and changing daylight cues. If you listen carefully in spring, you can hear the dandelions dancing to the call of nature's rhythm.

We know how to measure time, but we still question the nature of time. In 1687, Isaac Newton (1642–1727) stated that "absolute, true, and mathematical time, of itself, and from its own nature flows equably without regard to anything external." In 1905, Albert Einstein (1879–1955) disagreed.

In that year Einstein published his famous special theory of relativity. Among other things, it dealt with the effects of motion at speeds close to that of light. The velocity of light, which is approximately 186,000 miles per second, is, according to Einstein's theory, the fastest speed that any object can approach. At speeds that are significant proportions of the velocity of light, things do not happen as common sense tells us they should.

The theories of Einstein extended the theories of Isaac Newton to situations that in Newton's time were beyond comprehension. The results of Einstein's mathematical investigations of these situations are often beyond today's common sense.

The example most often used to explain Einstein's relativity time is that of the "twin paradox." In our imaginations we send twins to separate stars and then have them return to earth. The twins travel through space at speeds that are significant proportions of the velocity of light. The best places to send them are the stars Pollux and Castor, the two brightest stars in the constellation Gemini the Twins.

Twin A will travel at a speed three-fifths the velocity of light to Pollux, which is thirty-five light years away. Twin B will travel at a speed of four-fifths the velocity of light to Castor, which is forty-five light years away. The twins leave at the same time and they are waved off by their cousin, who was born on the same day as the twins.

At first glance, this seems like a straightforward problem. The round-trip for Twin A is seventy light years. Traveling at a speed of three-fifths that of light, the trip should take 116 years. The round-trip of Twin B is ninety light years. Traveling at a speed of four-fifths that of light, this trip should take only 112 years. This is how much time an earth clock would record for the trips of the astronaut twins. The clocks within the moving spacecraft would not agree.

Einstein's special theory of relativity predicts time will slow down in a spacecraft traveling at speeds close to that of light. In Spacecraft A, traveling at three-fifths the velocity of light, time will proceed at four-fifths the rate on earth. A clock in Spacecraft A will record a trip of 92 years, not 116 years. In Spacecraft B, traveling at four-fifths the velocity of light, time will proceed at three-fifths the rate on earth. A clock in Spacecraft B will record a trip of 67 years, not 112 years.

Here is where the paradox appears. The aging of the human body is similar to the working of a clock. If the clocks in a fast traveling spacecraft run at a slower rate, the human body will age at a slower rate. The returning astronaut twins will therefore be different ages because their bodies will have aged at different rates during the trip.

If the astronaut twins left the earth on the day of their birth, then Twin B would return to earth 112 earth years later at age 67, and Twin A would return four earth years later than Twin B at age 92. Twin B would be 71 (67 + 4) years old by this time. If the twins' cousin were there to meet the second returning twin, the cousin would be 116 years old. They all began their adventures at the same age, but they all finished at different ages.

If this does not make sense to you, it is because you are thinking in terms of Newton's description of time as flowing "equably without regard to anything external." Our everyday experiences are best explained and understood in Newton's terms. But experiments show Einstein's description of time to be closer to the truth. Highly accurate atomic clocks at the top of tall buildings run slower than atomic clocks at ground level. And atomic clocks in fast moving aircraft move slower than atomic clocks at rest on the surface of the earth.

What would happen if we were able to travel at the speed of light? If we could travel at the speed of light, time would stop and we would not age. We would be immortal like the Gemini twin, Pollux. Alas, since travel at the speed of light seems to be impossible, we are destined to be like Castor, the other, mortal Gemini twin.

5

Horizon Astronomy

They come from all over the world. By automobile or tour bus, it is a trip of less than one hundred miles from central London. The ride is made more pleasant by the beautiful rural countryside of rolling hills, but the anticipation still increases as the miles pass. At Amesbury they turn north after going three-quarters of the way around the "roundabout." One more small hill and they finally see it sitting alone in time and place. Stonehenge is a mecca for the curious and those who are interested in horizon astronomy.

Stonehenge is also disappointing. In reality it somehow appears physically smaller than it appears in one's expanded imagination. It is still impressive in spite of this.

Stonehenge is a three-dimensional rock and dirt puzzle whose origin and purpose have challenged the best of amateur and professional archaeological sleuths. British antiquarian William Stukeley (1687–1765) is credited with the first discovery of an astronomical nature at Stonehenge. In his 1740 publication simply entitled *Stonehenge,* he suggested the causeway leading from the central Stonehenge structure was pointed toward the rising point of the summer solstice sun in the northeast.

It was another Englishman, an astronomer named Gerald Hawkins, who created the later twentieth-century Stonehenge mystique. He turned the site into one of the most popular tourist attractions in England, second only to the Tower of London. Hawkins excited the world in 1963 by convincingly showing, with the assistance of a computer, that Stonehenge was, among other things, an astronomical calendar. His book, *Stonehenge Decoded,* was a bestseller, and his investigations of Stonehenge became the subject of a successful television documentary. Since that time Stonehenge has appeared on T-shirts, coffee mugs, postcards, and as the background for advertisements of commercial products of dubious connection to the site.

The work at Stonehenge was not finished with Hawkins's discoveries. There was still the question of "why?" Why would anyone build a massive stone calendar to mark the changing horizon appearances of the sun and moon? I believe I know the answer to this mystery. The "why" of Stonehenge is found in the plan of its earliest construction and a single stone—the Heel Stone.

Everyone who has ever been to Stonehenge has looked at the famous Heel Stone. This is the stone over which the sun is seen to rise on the first day of summer, the day of the summer solstice. The Heel Stone is a large upright rock that is stationed one hundred feet away from the main Neolithic stone and earth structure of Stonehenge. It was set in position to act as a marker for the rising summer solstice sun as seen from the center of the Stonehenge circle.

Many Stonehenge visitors, upon returning home and looking at their souvenir photographs, have probably remarked, mentally if not verbally, that the Heel Stone seems to possess facial features. This is a point that will usually elicit little argument. The turned-down mouth is quite prominent, as are the eye indentations and the slightly protruding nose. Of more fundamental importance is the question of whether the face is there by chance or by design.

If the face is there by chance, there is little to be learned from it. If the face is there by design, it may be possible the face can finally unravel the "why" of Stonehenge. How can we decide today, thousands of years after the fact, if the Heel Stone was chosen and placed as it was, *because* of the facial characteristics that it bears? We can't. We have to look at the Heel Stone in relation to the other stones and the history of the monument for an answer.

The construction of Stonehenge was begun around 2750 B.C., according to radioactive carbon-14 dating methods. It was built by an agricultural people known as the Windmill Hill People, whose name is derived from their burial mounds on nearby Windmill Hill. The Windmill Hill People are also thought to be responsible for the construction of Woodhenge, a wooden Stonehenge-type structure, and Silbury Hill, the largest prehistoric mound in Europe.

Stonehenge I, the earliest stage in the thousand-year history of construction and renovation at the site, consisted of a large circular ditch and embankment 350 feet in diameter that was left open to the northeast. It is speculated that two large upright stones were placed slightly inward of the entrance to the circle. The Heel Stone was placed approximately one hundred feet up the avenue or causeway that leads from the northeast opening. These three stones seem to be the only large stones used at Stonehenge I. Four smaller stones, called station stones, are also thought to be part of the initial construction. A narrow ditch was dug around the Heel Stone and later filled in with chalk to inhibit plant growth. No evidence of ditches has been found around the positions of the other large Stonehenge I stones. In fact, the two large entrance stones are no longer in place. Only the Heel Stone remains of the three original stones.

If we can assume that the purpose of the building of Stonehenge remained the same throughout the different periods of construction, then the original outline of the site was most important. The following years served only to improve and elaborate upon the original design in order to achieve that purpose, but not to significantly change it. The Heel Stone is the key to that purpose.

The thirty-five ton Heel Stone is twenty feet long, eight feet wide, and seven feet thick. It is buried four feet into the ground. But it was not just buried anywhere. It was put into the position required for the summer solstice sun to be seen to rise over it as viewed from the circular earth enclosure. Changes in the obliquity of the ecliptic and the subsequent tilting of the Heel Stone make it difficult to provide precise statements concerning the rising point of the sun on the Stonehenge horizon at the time of the building of Stonehenge I. The sun may have risen slightly to the left of the Heel Stone. It may not have been important for the early builders to have seen the summer solstice sun rise directly behind or over the top of the Heel Stone. We will unfortunately never know. The intent, however, is plain: The builders of Stonehenge I had the position of summer solstice sunrise point in mind when they placed the Heel Stone in position.

The Heel Stone was also placed so that the mute stone face could be viewed from the center of the circular formation. At first glance this view of the face is disappointing. You would expect the Heel Stone face to be staring directly into the center of the Stonehenge formation. It does not. It points to the right as seen from the center of the Stonehenge structure. This is significant because the face is actually looking almost directly south. This is the position of the sun at local noon when it crosses the meridian, and the position of the sun when it attains its highest daily altitude.

The triangular shape of the face causes one side of the face to be illuminated in the morning hours, and the other side of the face to be illuminated in the afternoon hours. At noon, the entire front of the face is illuminated. The Heel Stone face acts as a crude form of sundial.

If it is granted that the Heel Stone was chosen because of its resemblance to a face, and if it is further granted that it was purposefully placed as it is seen today, then an important question arises. What did the Heel Stone represent to the builders of Stonehenge I? The answer is obvious. The Heel Stone was used as a symbol or representation of a sun god. Imagine the excitement that would be generated by the view of the sun rising over the face of the Stonehenge I sun god on the day that the sun is in the sky for the longest period of time. But this was not the only day of the year that the Heel Stone served a function. Its sundial face served to tell morning, noon, and afternoon sun time on every day the sun produced a shadow.

Many civilizations have superimposed a face on the sun to indicate a deity. Christian art, for example, represents holy people with halos that are a form of sun disk. The custom of placing the sun behind or above a head to indicate divinity did not originate with Christianity, however. It goes back to the Egyptians and farther.

So what was the purpose of Stonehenge? We really can't say with certainty. We can only look at what is left and make inferences. Based on observations of the Heel Stone, a case can be presented that Stonehenge may have been a temple that took the form of a stone solar calendar that was dedicated to, or at least commemorated, an all-important sun god.

The people who built Stonehenge I had an undeniable knowledge of astronomy and mechanics. These were the practical arts, so they were developed first. The people who built Stonehenge were engineers, not artists. The representational arts came later. When the builders of Stonehenge I felt the need for a permanent temple to worship or honor their sun god, they may have used what was available. In this case it included a huge, multi-ton slab of rock that strongly suggested a face—the face of the sun god.

Their need for a sun god temple may have been derived from trade contacts with other civilizations. Perhaps they had heard stories or seen drawings of gods carved in stone. Perhaps out of pride they wished to duplicate these known aspects of foreign cultures.

The Stonehenge of today required several hundred more years of construction. Stonehenge II, which came into existence around 2000 B.C., required the transport of eighty four-ton bluestones over a land and water distance of 250 miles from the Prescelly Mountains of southern Wales. These stones were probably meant to form a double circle in the center of the circular earth structure. The circles were never completed on the west side.

Stonehenge III was a radical reconstruction of Stonehenge II. The two bluestone circles were dismantled and replaced by eighty sarsen stones. The much larger, sandstone sarsen stones weighed up to fifty tons and had to be brought from Marlborough Downs, a distance of twenty miles to the north. These stones were placed in a lintelled circle and a horseshoe of trilithons.

The dismantled bluestones were used again to form a circle and horseshoe within the sarsen circle and horseshoe. Stonehenge III was completed approximately one thousand years after Stonehenge I was begun. With the exception of a few stones that have disappeared and a number that have toppled from their original positions, this is the Stonehenge of today. This is the Stonehenge that Hawkins discovered with the alignments in the directions of the rising and setting sun and moon at the winter and summer solstices.

It seems significant that the Heel Stone endured in its original position during this long period of extensive modification. It also seems significant that it is the only stone of the seventy-six sarsen stones at the site that has not been hammered and pounded into a rectangular shape.

There have always been stargazers. People from all cultures down through the ages have been fascinated by the objects of the night sky, and the special star of the day sky. And there have always been people who have used astronomy to fulfill a need. Stonehenge is unique only in the sense that it is the world's most famous example of the use of horizon astronomy to fulfill a need. But Stonehenge is just one of a number of structures in the world that incorporate horizon alignments of an astronomical nature. There are other structures aligned to commemorate significant days of the year.

As time goes by, more and more is being learned about the use of astronomy by ancient cultures. Or, you could also truthfully say, the astronomy left behind by ancient cultures is teaching us more and more about ancient cultures. This has led to a renewed interest and appreciation of ancient cultures and a renewed interest in the discovery, reconstruction, and preservation of the monuments and structures they left behind. An example of this can be found thirty miles north of Dublin, Ireland.

Newgrange is a passage grave structure. It was built around 3000 B.C. by farming people who, like the original builders of Stonehenge, did not have the advantages offered by the wheel or metal tools to assist them in their work. And, yet, they transported stones weighing tons over distances of many miles to build what they believed to be a proper burial structure for the ashes of their dead. The Newgrange tomb consists of a passageway embedded in a football-field-sized mound constructed of stone and sod. The underground, sixty-foot-long, narrow, upsloping passageway leads to an end chamber and two side chambers. The three recesses each contain large bowl-shaped stones with shallow recesses on top. The bones and cremated remains of those interred there were placed in the bowls.

The people who built Newgrange were eventually forgotten. And so was their passage tomb structure. Nature took over. The tomb became camouflaged by the deterioration of the mound and the growth of trees and scrub. It became just one of several mounds dotting the banks of the River Boyne. Newgrange was rediscovered in 1699. The stones of the structure were being hauled away for road building when a large spiral engraved stone and the entrance to the passage were uncovered. It became and remained a little visited tourist spot for the curious from then on.

Newgrange became a part of ancient astronomy in 1963 with the discovery of what is called the "roof box." It was a specially constructed opening above the entranceway to the tomb. The opening was carefully constructed to admit the light of the rising sun down the passage and into the back recesses of the tomb. But it only did this at one time of the year—the day of the winter solstice. For a total of seventeen minutes each year, the passageway and three burial chambers are illuminated by the rising winter solstice sun. Newgrange is another example of the use of horizon astronomy.

Why is the Newgrange passageway to the three burial chambers aligned with the rising winter solstice sun? We can only guess. No written records were left behind. The day of the winter solstice can be seen as a day of rebirth for the sun. Following the winter solstice, the days start to become longer, the noon altitude of the sun starts to become higher, and the sunrise and sunset points start to move northward. Perhaps this day also signified a day of rebirth or resurrection for the deceased whose cremated remains were placed within the burial mound. Admittedly, it's a guess.

The ancient Egyptians made the most varied use of horizon astronomy. The earliest and most astounding use of horizon astronomy by the Egyptians can be found in the orientation of the great pyramids of Giza just outside of Cairo. The pyramids are the only surviving members of the Seven Wonders of the Ancient World. The Great Pyramid of Khufu is the largest pyramid ever built. Constructed in 2680 B.C. to a height of 482 feet, the Great Pyramid of Khufu consists of more than 2.3 million limestone blocks, weighing 2.5 tons each. Each side of the pyramid's almost perfectly square thirteen-acre base measures 755 feet. The difference between the longest and shortest base length is less than eight inches. The same accuracy of measurement was achieved in the nearly perfect orientation of the sides of the pyramid base toward the north, south, east, and west cardinal points.

This orientation was achieved by an accurate measuring of the rising and setting points of a seasonal star in the northern sky. Halving the angle between these two horizon points provided a determination of the true north-south direction. The east-west direction would lie at right angles to this line. The pharaohs later reenacted this important alignment process in a symbolic ceremonial event known as the "stretching of the cord."

When the last of the pyramids was built, the use of an astronomical horizon alignment was not abandoned. The Great Temple of Amon-Re at Karnak, built by Ramses II in the thirteenth century B.C., has an astronomical horizon orientation. The Karnak temple complex, the largest temple complex in the world, has its main axis aligned to the southeast direction of the winter solstice sunrise.

The solstices and cardinal points were not the only directions used for temple alignment in ancient Egypt. The temple at Abu Simbel, located in the upper Nile River Valley, is aligned toward an alternative direction along the horizon with an astronomical connection.

Knowledge is power, and the Egyptian pharaoh Ramses II knew this long before the English philosopher Francis Bacon expressed it in the seventeenth century. Ramses II knew enough about psychology and astronomy to effectively take advantage of the knowledge equals power equation. He turned it into an action. The resultant temple hewn out of a rock cliff face is impressively fronted by four sixty-five-foot-tall seated figures of Ramses II. The statues of Ramses dwarf those of his standing family members.

Ramses knew about the use of psychological tie-in signals to indicate the existence of advantageous personal relationships. He knew that "who you know" was important. And he recognized body proximity and orientation as the least subtle of all psychological tie-in signals. This is why his life-sized statue was strategically placed within a grouping of the statues of three important Egyptian gods. And this is why the statues were placed at the end of the two-hundred-foot-long corridor leading to the sanctuary within the temple of Abu Simbel.

It has been suggested that the large temple at Abu Simbel was designed as it was, and built where it was, as a behavior modification device. Ramses meant to impress the never-happy Nubians of the upper Nile River Valley with his power. He probably intended to suppress their constant uprisings by demonstrating that he had powerful friends in high places, and that the consequences of continued political misbehavior could be severe.

To be sure that the Nubian leaders got the message, Ramses added a little astronomical extra to the temple. He had the temple oriented to face a particular point on the southeastern horizon. It was the point on the horizon above which the sun rose on October 22 and February 22.

When the sun rose on February 22 and October 22, it traveled straight down the two-hundred-foot-long corridor of the temple and into the sanctuary, directly illuminating the statue of Ramses II seated between the statues of two deities. These days were important to Ramses II: February 22 was his birthday and October 22 was the day he celebrated the thirtieth year of his reign. (His total reign lasted an incredible sixty-seven years.)

Time and sand buried the temple at Abu Simbel with its six-story-high statues on the front facade. It was only accidentally discovered in 1812. The construction of the High Aswan Dam in the 1960s threatened to bury the temple again, this time under the rising waters of Lake Nassar. The rock cliff-embedded temple was saved, however, by being dismantled and moved piece by piece to higher ground. The new location preserved the sunrise phenomenon. The result of the reconstruction is an even more impressive tourist attraction, because of the engineering feat involved in the dismantling and reconstruction.

The temple of Abu Simbel, like all temples and tombs in ancient Egypt, has another horizon connection. It is on the east bank of the Nile. All Egyptian temples were built on the east bank of the Nile because temples were for the living. The life-giving sun appeared each day above the eastern horizon.

Tombs, such as the great pyramids at Giza, were built on the west bank of the Nile because tombs were for the deceased, and the life-giving sun ended each day by disappearing below the western horizon.

An even more dramatic horizon astronomy effect than that

at Abu Simbel was achieved with the light of the setting sun at Chichen-Itzá, an abandoned Mayan city in the Yucatan jungle of Mexico. One of the central buildings of the site is the temple-pyramid dedicated to Kukulcan, the Feathered Serpent. The 78-foot-high structure is called El Castillo, meaning the castle, a name given to it by the Spanish conquerers of Mexico.

The time to be at Chichen-Itzá is the time approaching sunset on the day of the March or September equinox. The northern side of the pyramid will be completely in shadow except for the stone serpent's head at the bottom of the western balustrade of the ninety-one steep steps leading to the top of the pyramid.

The show begins when the equinox sun gets low enough for the corners of the nine terraces that form the stepped pyramid to cast triangular shadows on the stairway balustrade. Triangles start to appear at the top and move down to the serpent head at the bottom of the stairs. The interplay of light and shadow forms an image of an undulating serpent descending the stairway.

The Mayans obviously went to a lot of trouble to create an event for the equinoxes, and one that deserved comment!

The people who gather at Chichen-Itzá today to witne tl.e descending serpent are there for a different reason than those who gathered there on the equinoxes five centuries or more ago. The once brightly colored and decorated pyramid of Kukulcan, the Feathered Serpent, was not part of a Mayan Disneyland. It provided another type of entertainment. It was a religious structure dedicated to a god who was thought to be pleased by human sacrifice. The temple at the top of the pyramid may have been used as an altar where the beating hearts of sacrificial victims were torn from their bodies. The skin of the sacrificed victims was then removed to be worn by others as a further part of the religious ceremony.

The El Castillo pyramid of Chichen-Itzá has a few other inter-esting astronomical connections, but nothing to rival the great diamondback rattlesnake show of the equinoxes. The pyramid has 91 steps on each side of its four sides. Four times 91 equals 364.

If you add the step which is shared by all the stairways at the top of the pyramid, you have steps totaling the number of days in the solar year.

Another of the famous structures at Chichen-Itzá is a building called El Caracol by the Spanish conquerers. It was named "the snail" because of its unusual, but now mostly ruined, cylindrically shaped tower. The surviving windows of El Caracol are astronomically aligned with the most northerly and southerly horizon rising and setting points of Venus. The planet Venus was a very important celestial object to the Mayans because of their concern with calendar systems. The Mayans had developed an extremely accurate calendrical system based upon the movement of Venus with respect to the sun (see pp. 105–106). Their word for Venus meant "great star."

The Maya achieved the highest culture in pre-Columbian America, while at the same time they ritually sacrificed thousands by cutting out their beating hearts. They had the most advanced calendar system in the world, while attributing eclipses to the voracious appetite of ants. The Mayan civilization lasted for two thousand years and then, for reasons still unknown, it collapsed. They abandoned their great cities and left us with one of the great unsolved mysteries of anthropology.

Charles Dickens's opening lines to *A Tale of Two Cities* could have just as well have been used to describe the Yucatan of Mexico at the time of the Maya: "It was the best of times, it was the worst of times, it was the age of wisdom, it was the age of foolishness, it was the epoch of belief, it was the epoch of incredulity."

The Mayan civilization is thought to have begun in the highlands of Guatemala around the fifteenth century B.C. and spread down on to the giant Yucatan peninsula that separates the Gulf of Mexico from the Caribbean Sea. The Yucatan area reached its cultural height during the eight centuries prior to about A.D. 1000. This is what is called the Classic period of pyramid and temple building. The decline of the Maya began with the infusion of the warrior-dominated Toltec culture from the central plateau of Mexico. They came from another city with astronomical alignments.

Teotihuacan, just thirty miles northeast of Mexico City, was the largest of the ancient cities of Mesoamerica. Indeed, its estimated population of 125,000 people made it one of the largest cities in the world of its time. The remains of the city are dominated by two giant pyramids that are astronomical in name only. The Pyramid of the Moon at the end of the Street of the Dead is 137 feet high. The Pyramid of the Sun, farther down the street, stands more than 200 feet high. In fact, Teotihuacan was not influenced by the sun and moon. It was influenced by a small grouping of stars.

The Pleiades, a conspicuous open star cluster in the constellation Taurus the Bull, seems to have been a dominant cosmic force in the minds of the designers of Teotihuacan. The city is thought to have been planned so that its main avenue was perpendicular to a baseline that pointed to the horizon setting position of the Pleiades.

The helical rising of the Pleiades in the eastern sky announced the first of two days of the year that the sun would cross directly overhead. On these days, the zenith noon sun did not cast a shadow. It was also on these days that the sun god descended to the earth, bringing a welcome rain caused by the heat of the sun's vertical rays. These were important days to an agricultural community like Teotihuacan. They were also important days for calibrating their calendar, since the summer solstice occurred midway between the days of the sun's two zenith passages.

The proposed astronomical orientation of Teotihuacan was followed as a tradition in the construction of new Mesoamerican cities for as long as fifteen hundred years. By this time, the original reasoning may have been lost, since the earth's motion of precession had changed the Pleiades helical rising days and had also moved their horizon rising and setting points. Traditions often persist beyond the period of memory of the reasons for their inception.

The most famous building complex with an astronomical horizon orientation in the Eastern Hemisphere is found in the "lost city" of Angor Wat in Cambodia, which is now called Kampuchea. The temples of Angor Wat, now partly reclaimed by the encroaching jungle and suffering the effects of war and neglect, were built in the twelfth century by the Khmers, a people to whom astronomy was a sacred science.

Angor Wat was dedicated to the worship of Vishnu, a solar-related Hindu god. It is not surprising, therefore, that the temple complex at Angor Wat was built with solar and lunar orientations.

An observer standing in front of the western entrance gate, for instance, will see the spring equinox sun rise directly over the top of the central tower of Angor Wat. Winter and summer solstice alignments are also observed from this western gate. Angor Wat has not been fully researched, but it seems solar and lunar cycles, as well as calendrical and cosmological time cycles, are incorporated into the architecture of the temple complex. The same must be true of other uninvestigated temples in Southeast Asia.

Horizon astronomy involves more than orientations toward particular points on the horizon. It involves the determining of directions on the horizon and the determining of positions on earth. There are particular stars in the night sky that have proved of practical value to all the trading and exploring cultures that have sought the horizon. The stars of the night sky have never served a more practical purpose than when they led the way for explorers to circumnavigate the globe.

Without a knowledge of the stars, the mariners of yesterday could never have widened the known horizons of the sea. They could never have traced their voyages of exploration, and they could never have provided "signposts" for the traders who rode in their wake.

The most important piece of information to a navigator out of sight of land on the open sea is his position. It is from this information that the quickest and safest course can be plotted to a desired destination. Before the age of radio, satellites, and inertial navigation systems, the navigator had to rely on the stars to determine a ship's position. Navigation by the stars required a familiarity with a series of bright stars distributed over the entire sky. These were the stars that are part of the present grouping of fifty-seven stars that comprise the navigational stars.

Every position on the earth can be uniquely described by two coordinates. These are the coordinates of latitude and longitude. To know your latitude and longitude is to know where you are.

Latitude is an angular measure of position north and south of the terrestrial equator. In the Northern Hemisphere, latitude is the easiest positional information to determine, since the altitude of Polaris above the north point on the horizon is equal to the latitude of the observer north of the terrestrial equator.

Polaris is the "north star," the star that is so close to the projection of the earth's axis of rotation into the night sky that it does not appear to move. It appears to remain at a fixed number of degrees above the horizon at all times for a given latitude position. As one moves to more northerly latitudes, Polaris moves progressively higher in the night sky. As one moves toward the equator, latitude zero degrees, Polaris moves closer to the northern horizon.

There is no equivalent of a north star in the Southern Hemisphere. This made open sea travel more difficult in the Southern Hemisphere and may be a factor in an explanation for the dominance of sea-faring cultures in the Northern Hemisphere.

While latitude determination was straightforward in the Northern Hemisphere, longitude determination was difficult in both hemispheres. Longitude is an angular measure of position east or west of a given north-to-south reference line on the earth. Determining longitude at sea was difficult because it required an accurate knowledge of time. The movement of the moon through the background stars had to provide the hands and face of the sea navigator's clock until the development of sea-faring mechanical clocks. The moon clock had serious limitations, including the obvious facts that it could not be seen on cloudy nights and it was not visible near the new phase.

Longitude was determined at sea by the measurement of the zenith angles of three stars at a given time. By determining the angular distance of a bright star from the point in the sky directly overhead, and knowing the time of the observation, the navigator was able to determine that a ship's position was somewhere on a given circle on the earth. Only if the ship were somewhere on this circle would the star appear at the measured angle from the zenith. Three zenith angles from three stars would provide three circles. The single point where all three circles intersected was the ship's calculated position.

The system sounds deceptively easy. But it is another case of "garbage in, garbage out." The discovery of America by Christopher Columbus illustrates this point. The truth of the matter is that Columbus discovered America because he believed in some bad astronomically determined information. He believed the earth was smaller than it actually is, and that Asia extended farther east than it actually does. He died thinking the same things.

The first determination of the size of the earth was performed by the Greek astronomer Eratosthenes of Alexandria in the third century B.C. He knew that an obelisk in Syene did not cast a shadow on the first day of spring, but that one in Alexandria did. He reasoned this was due to the curvature of the earth, and further, that a simple proportion existed between what he knew and what he wanted to know, which was the size of the earth. His calculations gave the earth a circumference of 28,700 miles, a value too large by 3,900 miles.

Eratosthenes's measurement meant that one degree of longitude at the equator was equal to 80 nautical miles. Other geographers arrived at other answers by other means. Ptolemy of Alexandria in the second century A.D. arrived at one degree being equal to 50 miles. Alfragan in the ninth century estimated the length of a degree as 66 miles. Columbus based his voyage on 45 miles per degree of longitude. The correct answer is 69 miles per degree at the equator.

This error was multiplied by the error assumed for the eastward extent of the Asian continent from Europe. Columbus used a two-hundred-year-old estimate by Marco Polo of 244 degrees of longitude between Portugal and Japan. The correct answer is 150 degrees.

Columbus used the two inaccurate figures that provided the shortest distance to Japan, China, and India. He anticipated a total voyage of 68 degrees of longitude. If each degree required a voyage of 45 miles, the entire trip would have been completed in 3,060 miles. As he sailed west from Spain, he faced a voyage of 210 degrees of 69 miles each. Columbus faced an actual voyage of 14,450 miles. He was extremely lucky America got in the way! Columbus proved that two wrongs sometimes make a "sight." He didn't realize his mistakes, however, and that is why the natives of America were called Indians.

In order to use the stars for accurate navigation, it is first of all necessary to have very accurate star charts of star positions and very accurate terrestrial maps of earth positions. And before the development of nautical clocks it was also important to know how the moon moved through the background stars with great precision. In 1675, Charles II of England ordered the establishment of his royal observatory in Greenwich, just a few miles down the Thames River from London, for "finding out the longitude of places for perfecting navigation and astronomy" and "for the observer's habitation and a little for pompe."

Some of the most notable individuals in the history of astronomy, including Edmund Halley (1656–1742), served as England's Astronomer Royal at the Greenwich Observatory. The observatory was designed by the celebrated English architect and one-time astronomy professor at Oxford, Sir Christopher Wren. Wren is most famous as the architect of St. Paul's Cathedral in London.

Greenwich Observatory became the world's observatory on October 13, 1884. It was on that date that delegates from twenty-five countries attending an international conference in Washington, D.C., voted to adopt the meridian line passing through the center of a Greenwich Observatory telescope as the single prime meridian for all nations. The meridian of a particular location is the line between the north and south pole that passes through that location. The Greenwich meridian became the zero meridian from which all longitudes on earth were referenced.

The Greenwich Observatory was the reasonable choice for prime meridian since 65 percent of the ships in the world were already using navigational charts based on the Greenwich meridian. Before the 1884 decision, different countries used different lines for zero longitude on their navigational charts.

Horizon astronomy has served the political and economic interests of governments around the world. This has certainly been true in the past and it will probably prove to be true again in the future. It was especially true in the eighteenth century when Britain funded the expeditions of Captain James Cook.

Cook's three expeditions to the South Pacific were billed as voyages of discovery and scientific advancement. His instructions were to search for a mythical southern continent, "terra australis," to search for a western entrance to a northwest passage across North America, and, finally, to add to the knowledge of distant lands already discovered.

In carrying out his mission, it was important for Cook to know where he was and where he had been. This was part of his scientific mission. Cook's scientific goals were to perfect the methods for determining time, and therefore longitude, at sea, and to observe the 1769 transit of Venus.

Transits of Venus are annular eclipses of the sun. They take place when Venus moves between the earth and sun. A Venus transit does not take place every time Venus passes between the earth and the sun. Sometimes Venus will pass above the sun as seen from earth and sometimes it will pass below. Venus is not large enough to significantly obscure the light from the sun, as does the moon when it moves in front of the sun. It would take 116 planets the size of Venus side by side just to stretch across the diameter of the solar disk.

Transits of Venus are also very rare. They occur in pairs separated by eight years, but the pairs are separated by more than a century. No astronomer alive today has observed a transit of Venus. This is because the last two transits occurred in 1874 and 1882. The next two will occur in the years 2004 and 2012.

Accurate observations of the times of the beginning and end of a Venus transit from different points on the earth can lead to a value for the distance between the earth and the sun. Cook and the scientific members of his crew observed the June 3, 1769, transit of Venus from the vicinity of Tahiti. The results were disappointing, in spite of the excellent weather conditions. The problem was in the difficulty of timing the exact moment of the transit events. An error of just a few seconds was enough to make the results unreliable.

Cook was not just a competent sea captain and navigator; he was also a self-taught astronomer whose skills were recognized by the Royal Society. This probably explains the number of other phenomena observed and recorded during his voyages. Many of the observations required his ships to be at a given location at a given time.

On November 9, 1769, Cook observed the transit of Mercury across the sun from "Mercury Bay" in New Zealand. Transits of Mercury are as rare as transits of Venus. A total eclipse of the sun was observed from Tonga in June 1777, and another was observed in December of the same year from Christmas Island. Some of Cook's observations of astronomical phenomena were not location-dependent. These included comets and the aurora seen from the icy waters of the Antarctic and Arctic.

Cook's most important celestial observations were those of

the moon and stars. These observations allowed him to determine the longitude and latitude location of his discoveries. This in turn increased the accuracy of his charts. And it was the charts that allowed Cook and those who followed him to return. Cook's observations of the moon and stars also allowed him to test the accuracy of the experimental sea chronometers he carried on board. The development of mechanical sea-faring chronometers would eventually eliminate the need for lunar observations as a means of telling time.

Cook's expeditions into the Pacific produced some very practical results. He trained a generation of sailors and officers for duty in the Pacific. But they returned using horizon astronomy techniques as traders, *not* as explorers. The political and economic motives behind Cook's voyages of scientific discovery finally surfaced. With the loss of Britain's colonies in North America, the timing could not have been better. The sun was soon never to set on the British Empire. There are interesting parallels between Cook's voyages of discovery and the early United States space program.

The system of using stars to determine earth position is still part of the training of navigators, but it has been relegated to a back-up system for emergencies. This, however, does not mean that the bright navigational stars have lost their importance. The sea explorers of yesterday have been replaced by the robot space explorers of today. But the navigation stars are still widening the known horizons, only now it is the horizons of space rather than the horizons of the sea.

Spacecraft leaving the shores of earth need to know their position in the open seas of space. It is from this information that the quickest and safest course can be plotted to a desired rendezvous within the solar system. Solar system navigation has an added problem in that the destinations are all moving in orbits around the sun or another body. If there were no stars, there would have been no exploration of the solar system beyond the reach of the telescope. There would have been no horizon to show the way.

Horizon astronomy is not the exclusive domain of humanity. Humans are not the only creatures on earth to use horizon astronomy for navigation. Research has shown that migratory birds have also mastered the art of celestial navigation. By observing the rotation of the night sky they are able to locate Polaris and the direction of north on the horizon.

Birds and humans have problems finding stars on cloudy nights, but birds apparently have the added ability to determine directions by the earth's magnetic field. There is also evidence that they can use ultraviolet light and polarized light as navigational guides. A "bird brain" is a powerful navigational device that allows migratory expeditions to be accomplished over thousands of miles.

If you extend your definition of an astronomer even further beyond the human realm, then you will find that astronomers are literally everywhere. You never know where you are going to run into an astronomer because there are astronomers in the insect world. You can even find them at the seashore.

Now you are probably saying to yourself, "Oh, here comes one of those tired old jokes, like the T-shirt that proclaims the wearer to be a 'beach astronomer' because he or she observes heavenly bodies." This is no joke; the seashores of the world are infested with naked-eye astronomers. The variety at the New Jersey shore is known to the academic world as *Talorchestia megealop-thalmia* and *Talorchestra longeicornis.* They are more commonly known as beach fleas.

Beach fleas are small animals that live in the moist sandy sections of a beach that are not constantly battered by waves. They live there because that is where the good life is, as far as they are concerned. The nearby waves constantly deposit a chow line of delicious morsels of algae, seafood carrion, and assorted debris for their delightful dining.

During the earlier part of the day, beach fleas stay buried

in the sand to avoid the drying effects of the sun. They only begin to step out for dining in the late afternoon. When they do, they use an astronomical orientation method to get from one dining place to another and home again.

This requires beach fleas to have an internal physiological clock to compensate for the changing angle of the sun during the day, and changes in the length of day with the seasons. Experiments indicate the beach fleas' biological clock is temperature controlled to adjust to the seasonal changes. They use their eyes for the astronomical clues that allow them to travel rather than using some other sensing device. Blindfolded beach fleas get lost.

How does a low-life beach flea learn all of this astronomy? The answer seems to be that it is an inherited trait. A beach flea born on a western coast will take off in the wrong direction looking for the tide line if transported to an eastern cost. The beach fleas' genetic adaption to a particular location makes their ability completely worthless and potentially lethal at a totally different location.

Humans are still the best horizon astronomers.

6

Crystal Ball Astronomy

The future is more interesting then the past because it is unknown. It's mysterious. But humans like to know. We are curious, so we try to take all the mystery out of the future by attempting to figure out just what the future will be.

We have an increasing desire to predict the future. The feeling is that if we can accurately predict the future, we can take advantage of this knowledge. Knowing the future, it may even be possible to change the future before it gets here.

The future is looked for in signs—signs in the present and signs in the past. Those who find the true signs and interpret them correctly have an edge on those who do not. They are the winners, the true prophets. Those who follow false signs, or incorrectly interpret the true signs, are destined to be losers. They are the false prophets.

The sky definitely contains signs that forecast the future. Crystal ball astronomy has its roots in the attempt to use these signs.

Meteorology is one of the successful outgrowths of crystal ball astronomy. The word "meteorology" was originally used to mean the study of all signs derived from phenomena above the earth. It originally also dealt with astronomical phenomena. The word acquired the more restricted use it has today with our evolving knowledge of the location of the observed phenomena of the universe. Today meteorology refers exclusively to the study of the phenomena of the atmosphere, especially weather and weather conditions. The word "meteorology" is a linguistic fossil of an earlier time in the history of interpreting signs in the heavens.

It is possible to observe the meteorological signs in the heavens and make predictions of the future without any knowledge of the physical dynamics of the atmosphere causing the signs. It has been done in the past. Sky watchers of the past learned some of the signs and successfully arrived at their correct interpretations. Some of the signs have probably been forgotten, but some have been remembered.

The ancient Greeks successfully used the constellation Cancer the Crab to predict changes in the weather. On the darkest and clearest of moonless nights, it is possible to see a hazy patch of light close to the center of the inverted "Y" of the stars that form this faintest of zodiac constellations. This hazy patch becomes a beautiful cluster of several dozen sparkling stars when viewed through binoculars or a telescope.

If this hazy patch of light could not be seen on what appeared to be a clear night, then the ancient Greeks knew there was a good possibility the weather was changing. The cluster's disappearance was due to the presence of high altitude cirrus clouds. If cirrus clouds thicken to become cirrostratus clouds, then rain is a possibility.

One of today's best known predictive meteorological signs is remembered as a maritime mnemonic rhyme:

> Red sky at night, sailor's delight.
> Red sky in the morning, sailor take warning.

Even a landlubber who gets seasick taking a bath can understand the meaning of this rhyme, although the reasoning behind the forecast may remain elusive.

A fiery, western "red sky at night" during sunset arises from the setting sun lighting the undersides of high clouds. With predominately westerly winds at cloud level, the red sky forecasts dry, cloudless air coming into the region from the west. The sailor can take delight in having storm-free weather unitl at least the next day. A dull red sunset blotted by clouds predicts the opposite.

A fiery, eastern "red sky in the morning" during sunrise is caused by the rising sun again lighting the undersides of high clouds. With predominately westerly winds at cloud level, this sign forecasts a warm front, with rain clouds already moving into the region. The sailor is forewarned.

There are other signs. A rainbowlike, circular halo surrounding the sun or the moon is likewise a meteorological sign. The halo will have an observed radius of 22 degrees, with a faint red tinge on the inside and a faint violet tinge on the outside. The halo is caused by hexagonal ice crystals moving into the sky by way of high altitude clouds. These clouds usually forecast the impending arrival of a warm-front rain system.

And, of course, there are the stars, the ultimate meteorological predictors. Any night that you can't see the stars is a night of bad weather.

While meteorology is one of the successful outgrowths of crystal ball astronomy, astrology is not.

Astronomy and astrology are often confused. Astronomy is the study of the radiation and motion of extraterrestrial objects. As such, it has little to do with astrology.

Astrology is the study of a supposed relationship between cosmic events and terrestrial events, as well as the individual consciousness. It is an attempt to use macrocosmic phenomena to predict microcosmic events. Actually, "predict" is too strong a word.

Astrology does not deal with cause and effect relationships, but rather with the possibilities or probabilities that events may occur in certain places at certain times. The motto of the astrologer is: "*Astra inclinant, non necessitant*—The planets incline, they do not determine."

Astrology began in ancient Babylonia as a state concern that dealt with national and world events. Given that its practitioners needed at least some celestial observations to perform their duties, Babylonian state astrology also incorporated state astronomy. Attempts to use the stars to make forecasts involving large groups of people is called "mundane" or "natural" astrology.

The split between astronomy and astrology began with the Greeks. The ancient Greeks extended astrological forecasting to individuals. They developed a "judicial" or "natal" astrology which inferred an individual's personality characteristics based upon the arrangement of the sky at the time of birth. They also introduced "horary" astrology, with its horoscopes for day-to-day advice. Most people think of birth charts and horoscopes when they think of astrology.

It was Claudius Ptolemy, a second century A.D. Greek astronomer at the Library of Alexandria, who established the astrological tradition of signs, ascendants, houses, and aspects. It was he who codified the astrology now traditionally practiced in the West.

Astrology, as opposed to astronomy, is an indoor activity. It is performed with books of tables of planet, sun, and moon positions, and calculations based upon the timing of events. Natal astrology, for instance, begins with the casting of a birth chart. This is a reconstruction of the geocentric picture of the astronomical situation at the time of a person's birth.

The first thing an astrologer determines is an individual's sun sign. This is determined by which of the twelve sections of the zodiac the sun was in at the time of the person's birth. This is not the same as the constellations the sun was in at the time of birth. The sun signs consist of twelve, 30-degree-wide intervals, measured eastward along the ecliptic. The zodiac constellations consist of twelve traditional star patterns of varying size arranged along the ecliptic. Sun signs and zodiac constellations are often confused because the names of the signs and the constellations are the same except in two cases. The confusion is increased by the fact that they are also in the same order.

TABLE OF SIGNS AND CONSTELLATIONS

STARTING DATE	SUN SIGN	CONSTELLATION
March 21	Aries	Pisces
April 21	Taurus	Aries
May 21	Gemini	Taurus
June 22	Cancer	Gemini
July 23	Leo	Cancer
August 24	Virgo	Leo
September 23	Libra	Virgo
October 23	Scorpio★	Libra
November 23	Sagittarius	Scorpius★
December 22	Capricorn★	Sagittarius
January 21	Aquarius	Capricornus★
February 20	Pisces	Aquarius

The reason the sun signs and zodiac constellations do not match up is due to the precessional motion of the earth's axis of rotation. Precession is a clockwise movement of the rotational axis of the earth around the direction of its revolutionary axis. The precessing rotational axis of the earth sweeps out a cone in much the same way that a spinning top will do as it slows down just before falling over. Precession is the result of the earth's rotation and the differential gravitational forces of the sun and the moon acting upon the earth's equatorial bulge that is a result of the earth's rotation.

As the earth's axis precesses, the north celestial pole will move to the vicinity of different stars. This will produce different "north stars" over the 25,800-year precessional cycle. Today the pole point is approximately one degree away from Polaris. In the year 300 B.C., at the time of the building of the pyramids, Thuban (Alpha Draconis) was the north star. At that time the north celestial pole was four degrees away from Thuban. In the year A.D. 14,000, one-half of a precession cycle from now, the bright star Vega (Alpha Lyrae) will be the north star at an angular distance of five degrees from the north celestial pole point.

As the earth's axis of rotation changes its direction in space, the equatorial plane, which is at right angles to the rotational axis, will change its orientation with respect to the ecliptic plane. The clockwise precession of the rotational axis therefore causes the two node points, the positions of the equinoxes on the ecliptic, to slowly move westward. The westward progression of the March equinox along the ecliptic is called the "precession of the equinox."

The March equinox position will move 360 degrees along the ecliptic and through each of the zodiac constellations in 25,800 years. This amounts to a movement through one constellation every 2,150 years. Two thousand years ago in the time of Ptolemy the signs and constellations of the same name were aligned, but due to precession, the signs, which are measured eastward from the March equinox, have slipped back one constellation. The sky of the astrologer is the sky of two thousand years ago. It is not the sky of today. This bothers astronomers. It does not bother astrologers. They either don't know about it, or don't care about it.

The second most important supposed astrological influence at a person's birth is the ascendant. It is considered to be as important as an individual's sun sign. The ascendant is the sign rising in the east at the time of birth. Precession also has caused the ascendant to differ from the zodiac constellation that was rising in the east at the time of birth. The sun sign and ascendant are to astrology what heredity and environment, nature and nurture, are to psychology.

Houses and aspects are also a traditional influence in a complete astrological forecast. Houses are twelve equal regions of the celestial sphere, starting at the eastern horizon, that influence different areas of human life. A planet, the sun, or moon in a particular house on the birth chart affects the area of life represented by that house.

Certain aspects or angles formed between planets, the sun,

and moon, as seen from the earth, are considered important in determining the overall influence of celestial bodies. These influences can be beneficial or harmful. Since true aspects seldom occur, an "orb of influence" is allowed that varies from astrologer to astrologer.

TABLE OF ASPECTS

ASPECT	ANGLE (DEGREES)	SYMBOL	INFLUENCE
Conjunction	0	☌	Mixed
Sextile	60	★	Favorable
Quadrature	90	□	Unfavorable
Trine	120	△	Favorable
Opposition	180	☍	Unfavorable

The casting of a birth chart to show the astronomical situation at the time of an individual's birth is the easy part of astrology. The sky looked only one way at the time of a person's birth. The difficult part is providing a meaningful interpretation to the physical situation. The data provide more than one interpretation. This is the "art" of astrology. The astronomical situation sets the stage for the astrologer's improvisation. There are no hard and fast rules in astrological interpretation. Every astrologer is his or her own master. Every astrologer has his or her own set of talents.

There are some general rules in astrology as to the characteristics assigned to the signs of the zodiac, and it is fairly easy to see how they were derived. They are suggested by the names of the astronomical constellations that preceded the creation of the astrological signs. If there is any doubt about this, one has only to inspect the characteristics assigned to the planets discovered in modern times. The characteristics assigned to Uranus, Neptune, and Pluto were suggested by the names assigned to the planets by astronomers. The astronomers assigned the names with no thought at all to any astrological influence the new planets might have.

The question of how the stars, planets, moon, sun, and the angles between them supposedly influence earthly affairs has never been adequately answered by astrologers. Perhaps for good reason.

Astrology is bad science, but it is good business. This is why astronomy and astrology existed in a symbiotic relationship until the seventeenth century. Astronomy was supported by astrology. Many famous astronomers were employed simply because their knowledge of the sky allowed them to cast horoscopes. Good astronomers, however, were not always good astrologers. Galileo once predicted a long life for a visiting dignitary to the Florentine court. The visitor died two months later.

Johannes Kepler was one of the greatest theoretical mathematical astronomers. He had to be a theoretician: He suffered from congenital myopia and multiple vision. Kepler was also a lifelong advocate of astrology.

Kepler's first professional position was that of a mathematics instructor in Austria. One of the traditional side duties of the position was the publication of an annual calendar of astrological forecasts. In his first calendar attempt, Kepler forecast an extremely cold winter for 1595–96, and an invasion by the Turks. The winter was extremely cold and the Turks devastated Vienna on January 1, 1596. The forecast of astrological calendars and horoscopes provided Kepler, for the rest of his life, with survival money to supplement the insufficient income he received from his astronomy positions. Kepler described astrology as the "stepdaughter of astronomy." If so, Kepler was aware the stepdaughter was in the higher tax brackets.

Astronomy and astrology separated for good three hundred years ago during the scientific reign of Isaac Newton. It was then that mathematics and the scientific method were adopted by astronomers as the most productive procedures to follow. Astronomy after that attained a respectability that astrology did not.

In many ways astrology has had the last laugh. In spite of everything astronomers have tried, people still show an interest in astrological predictions. Astrology has a definite appeal, a definite charisma. Unlike astrologers, astronomers are at a real disadvantage when they make predictions. They aren't expected to be creative. Everyone expects them to be rational and totally accurate. And when they are accurate, it is not a news event. Only the unexpected astronomical event is news. And no one is able to predict the unexpected.

Astrologers have more fun than astronomers at predicting; hardly anyone ever takes them to account for their extremely poor success ratio. When they prove to be correct with one of their predictions, even when it is after the fact, it is made into a news event. The misses are politely forgotten.

Astrology is not a science. Because of this, astronomers don't like to be confused with astrologers. They view them as charlatans. The astrologers, of course, are laughing all the way to the bank. This attempt at disassociation may be an impossible task because the words "astronomy" and "astrology" are too closely related in sound and association with the night sky.

Astronomers are at the same time jealous of the notoriety astrologers have and the financial rewards astrologers can reap. The yearly number of books and articles on astrology is far greater than the number concerning astronomy. Astrology columns are syndicated to an estimated fifteen hundred newspapers. Practically every newspaper in the world has a daily astrology column, but very few have even a weekly astronomy column.

The human desire to feel of cosmic importance has over the millennia fostered an overt interest in astrology. Followers of astrology believe they must be an important part of the universe if the universe beyond the earth has a strong influence over their lives. To be ignored is to be unimportant. In light of this, it should not be surprising to find that astrological beliefs have become deeply embedded in modern folklore—in superstition. It is surprising to find that astrological beliefs are deeply embedded in other apparently unrelated areas of our lives. It is often so deeply embedded that origins and connections have become obscured. Astrology is more a part of our lives than we usually realize or are willing to admit. For instance, we start each day with an astrological belief.

Our days of the week are named in honor of the sun, moon, and naked-eye planets. These are the seven visible celestial objects that move among the stars of the sky. This naming scheme is not entirely obvious to English-speaking cultures because most of the English names for the days of the week are derived from unfamiliar Nordic gods that are associated with these celestial objects. The exceptions of course are Sunday, Monday, and Saturday, our English language days honoring the sun, moon, and Saturn.

So how about Tuesday, Wednesday, Thursday, and Friday? What celestial objects do they honor? The answer is fairly obvious if the names for these days of the week are translated into Spanish or French.

In Spanish, Tuesday is martes, Wednesday is miercoles, Thursday is jueves, and Friday is viernes. In French, Tuesday is mardi, Wednesday is mercredi, Thursday is jeudi, and Friday is vendredi. The planet associations are obvious. Tuesday honors Mars. Wednesday honors Mercury. Thursday honors Jupiter. And Friday honors Venus.

Our English language days honor the same planets, but not in the same order. The celestial objects honored on Wednesday and Thursday are reversed in English. The English Wednesday comes from Woden's day. Woden was the Nordic king of the gods, the equivalent of Jupiter. Our Woden's day (Wednesday) is therefore equivalent to the Spanish jueves and the French jeudi (Thursday).

The English Thursday is Thor's day. Thor was the son of Woden and the counterpart of Mercury. Our Thor's day (Thursday) is therefore equivalent to the Spanish miercoles and the French mercredi (Wednesday).

In English, Spanish, and French, Tuesday honors the god of war. Tuesday in English comes from Tiw's day. Tiw was the Nordic god of war. The Spanish and French names for this day of the week are martes and mardi, honoring Mars, the more familiar god of war.

Friday in English is derived from Freya's day. Freya was the Nordic goddess of love and beauty. The Spanish and French names for this same day of the week are viernes and vendredi, honoring Venus, the more familiar goddess of beauty.

The question now arises as to how the days were ordered. The answer is not found in our present understanding of the solar system. If it were, the days of the week would be Sunday (sun), Thursday (Mercury), Friday (Venus), Monday (moon), Tuesday (Mars), Wednesday (Jupiter), and Saturday (Saturn).

The ordering of the days of the week goes back to a time when it was assumed that the earth was at the center of the universe. The ordering of the solar system above the earth from farthest to closest was then believed to follow the sequence: Saturn, Jupiter, Mars, the sun, Venus, Mercury, and the moon. This order, if used, would give us a week that would follow the progression: Saturday, Wednesday, Tuesday, Sunday, Friday, Thursday, and Monday. This is not the correct sequence, but with it, and one more piece of information, it is possible to arrive at the correct sequence.

The other piece of information comes from Egyptian astrology. The Egyptians believed that one of the seven moving naked-eye objects ruled each of the twenty-four hours of the day. Each day was then named after the celestial object that ruled the first hour of the day.

If you started the first hour of a day with Saturn, the farthest celestial object from the earth in ancient astronomy, then Saturday would be the first day of the week. In order to finish the twenty-four-hour day, and arrive at the first hour of the next day, you would have to repeatedly move through all seven celestial objects three times, and then move ahead four more (21 + 4 = 25 hours = first hour of next day). The first hour of the next day would be ruled by the sun. In English, this is Sunday.

The determination of the celestial object ruling the first hour of the day following Sunday would require the same repeated movement through the seven celestial objects three more times, and the same addition of four more. This would identify the ruler of the first hour of the next day to be the moon, and this would produce Monday.

The celestial objects ruling the first hour of the succeeding days would be Mars, Mercury, Jupiter, and Venus. In Spanish and French this fixes the days of the week in their proper sequence, but in English it gives the sequence of Tuesday, Thursday, Wednesday, and Friday. Using the ancient construction of the universe and the astrological belief in the hours being ruled by the seven celestial objects, the wrong order is derived for the English days between Tuesday and Friday. How or why this happened, I don't know.

The impact of astrology is certainly "more than meets the eye." The well-known poem defining the attributes of children born on different days of the week reflects the planetary associations with the days of the week. The poem says, "Thursday's Child has far to go." This is the child of Mercury, the fleet-footed messenger god. "Friday's Child is loving and giving." This is the child of Venus, the goddess of beauty and love. "Saturday's Child has to work for its living." This is the child of Saturn, the god of time and agriculture. "But the child that's born on the Sabbath Day is fair and wise and good and gay." The Greek sun god was Apollo, who was the god of truth and a master poet, musician, and healer.

A belief in the influence of the moon on human behavior is a form of disguised astrology that is more often thought of as superstition. The oldest moon superstitions developed from early attempts to discover why and how things happened. This explains why so many of the moon superstitions relate to agricultural activities.

Many moon superstitions involve the waxing (growing) and waning (decreasing) phases of the moon. The moon is waxing from the time of a new moon until the time of a full moon. During this two-week period, the moon "grows" more and more illuminated each night until it finally appears as a full disk. The moon is waning from the time of a full moon until it goes back to the new moon phase. During this two-week period, the moon "decreases" as it becomes less illuminated each night until it finally disappears from sight.

According to superstition, sowing activities are generally to be performed during the waxing moon period and reaping activities during the waning moon period. Peas should be planted during the waxing phase. Onions, however, since they do not produce an above-ground yield, should be planted during the time of a waning moon. Fruit should be picked in the fall during a waning moon so the bruises will dry rather than rot. Firewood should also be cut during a waning moon for the same reason.

The waxing and waning rules of superstition have been extended to include other human activities beyond those of agriculture. For instance, if you cut your nails or hair during the time of a waxing moon, they will grow back more quickly than if you cut them during the waning moon. If you dig a hole in the ground during the time of a waxing moon and then try to fill it again, you will find the amount of dirt has "grown." You will be able to fill the hole and have dirt left over. If you dig the hole during the time of a waning moon, however, you will get the opposite result. The amount of dirt will have "decreased" and you will not have enough dirt to completely fill the hole.

Supposedly, pointing at the moon will bring misfortune, and reading in the light of the moon will weaken eyesight.

Also according to superstition, the age of the moon on the

day of the first snowfall, in terms of the number of days since the new moon, will foretell the number of snowfalls to be expected within the winter months.

The most terrifying moon superstition involves lycanthropy, the power of a human to transform into a night-stalking werewolf in search of human prey. This is a world-wide superstition that will sometimes use the fiercest animal of the region, such as a tiger, leopard, or bear, in place of the wolf. The Hollywood wolfman movies helped to popularize this supposed effect of the full moon.

It is a time-honored belief that there is a cause and effect relationship between the full moon and criminal and abnormal mental behavior. It was such an established belief in England two hundred years ago that it was considered a factor in the courts of law. The law recognized a difference between people who were chronically insane, for instance, and those who were temporary "lunaticks." The latter were susceptible to aberrant behavior by virtue of the existence of a full moon. They were treated more tolerantly by the courts.

How could the moon affect human behavior? Those who believe in the reality of the full moon's influence usually cite the tides as their evidence. The argument goes like this: Since human bodies are composed of great amounts of liquid (90 percent water), the gravitational influence of the moon should affect humans just as it affects the waters of the earth to produce tides twice each day. In other words, the moon creates tides within the human body and these tides affect behavior. Established lunar rhythms in lesser organisms of the animal kingdom are often used as supporting evidence for this theory.

Studies attempting to prove or disprove the influence of the full moon on human behavior have produced mixed results. For every piece of research that establishes a statistical relationship, there is another that cancels it out. The major difficulty in the statistical correlation research lies in its inability to reproduce the proposed correlations in separate studies.

The guilt of the moon has not been established beyond a reasonable doubt. The moon has been accused of behavior modification, but the only evidence presented to support the claim has been circumstantial. Part of the problem is due to a general misunderstanding of the moon's presence in our sky. Unfortunately, the maxim "out of sight is out of mind" applies to the moon.

The moon rises above our eastern horizon and sets below our western horizon everyday. The moon is in our sky every day whether we see it or not. We do not see the new moon because it is too close to the sun and only the "invisible," shadowed side is facing us. We do not see the new moon, but it is in our day sky from sunrise to sunset. The waxing and waning of the moon is likewise an illusion. Only the percentage of the lighted portion of the moon that is visible is increasing or decreasing. When a cresent moon is in the sky, a whole moon is in the sky, *not* just the lighted portion of the moon that is visible.

If tides within our bodies are the cause of the moon's influence on human behavior, then human behavior should experience two tides a day, every day, just as the oceans and waters of the earth. The higher tides experienced at the time of a full moon have a counterpart in the high tides experienced at the time of a new moon. And, yet, there are no superstitions about the effect of the new moon on human behavior. The tide theory holds little water because it is too simplistic an explanation. It conveniently ignores the facts.

It is my guess that the moon is innocent of the charges, but not of involvement. The moon is only an accessory, a celestial "Typhoid Mary." The real culprit of the full moon's supposed influence on human behavior is the sun. The modus operandi is moonlight, which is nothing more than sunlight reflected by the moon. We see the moon by this reflected sunlight, and it is this extension of day into night by extra light that causes the extra amount of daytime activity associated with the full moon. The light of the full moon is bright enough to cast shadows, and this quite simply extends the length of the day, encourages mobility, and increases the possibility of human interactions. On a moonless night, you are probably safer in a dark alley than you are in a well-lighted parking lot. On a full moon night even the alley is well lighted.

And, of course, there is comet astrology. Comets in general, and Halley's comet in particular, have had a long association with death and misfortune. Comets were seen as evil omens from heaven, harbingers of war, pestilence, disease, and famine. They were equally perceived in the past as announcements of the dethronement of rulers and the dissolution of kingdoms.

William Shakespeare's lines from "Julius Caesar" summarize the comet political wisdom of the ages:

> When beggars die, there are no comets seen:
> The heavens themselves blaze forth the death of princes.

The most famous of the "deaths of princes" associated with Halley's comet was that of King Harold of England. Halley's comet appeared in the spring of 1066, and Harold was informed that it was an evil omen. The prophesy proved correct. William the Conquerer invaded England and Harold was killed in October 1066 at the celebrated Battle of Hastings. The famous two-hundred-foot-long Bayeux Tapestry graphically portrayed the history of these events, and produced the earliest known representation of Halley's comet by actual comet observers.

The tradition of Halley's comet as a bad omen goes further back than Harold. Its 12 B.C. visit was believed to forecast the demise of the Roman General Marcus Vipsanius Agrippa, who died soon after the comet's appearance.

The Jewish historian Flavius Josephus described the A.D. 66 appearance of the comet as a "star resembling a sword" that was seen over Jerusalem. Perhaps the Jews were hoping to use the comet to help overthrow their Roman subjugators when they rose in rebellion that same year. Unfortunately, the omen worked the other way, and Jerusalem and the Temple were destroyed.

The comet reappeared in 451 and this time it brought bad news for the infamous Attila the Hun who was at the height of his power. He was defeated in Gaul and forced to withdraw. He died two years later of a nose bleed suffered while celebrating his marriage.

Halley's comet enjoyed a reputation that would have been envied by the savage Attila during its 1456 return engagement. The dreaded Turks were threatening to invade Europe, and astrologers were predicting an imminent pestilence, famine, or some equally horrendous catastrophe. Pope Calixtus III supposedly attempted to counter the multiple threat to Christianity by ex-communicating the comet, ordering the ringing of church bells at noon, and offering the noon prayer, "God save us from the Devil, the Turk, and the comet." The strategy worked. The Turks were defeated, and the comet went away. But the noon church bells, and now their modern firehouse counterpart, have continued to ring into the twentieth century.

The superstitious belief that comets were portents of misfortune was supposedly put to rest in the eighteenth century with Edmund Halley's discovery that comets were ordinary members of the solar system that were only visible when their highly elliptical orbits brought them close to the sun, however. There are two modern astronomers who would seriously caution against taking the sup-posed harmful effects of comets too lightly. British astronomers Sir Fred Hoyle and Chandra Wickramasinghe have initiated the study of exopathology. Or should it be called astropathology? They are betting there is an element of truth in the comet superstition.

The diseases-from-space hypothesis proposes that comets are the teeming abode of disease-causing microorganisms. These extra-terrestrial life-forms arrive in the inner solar system as part of the tail of a visiting comet. When the earth intersects the space where the comet has passed, the microscopic extraterrestrial viruses drift down to earth shouting, "Take me to your leaders."

These invaders from space then stake out a claim on the cells of the human body, and without the annoyance of counteracting antibodies, they begin a population explosion. What follows is called an epidemic.

Nothing in the Hoyle-Wickramasinghe diseases-from-space hypothesis is implausible. Comets are known to contain carbon and the elements essential for the development of life. It is possible that the heat from radioactive atoms in the frozen nucleus of a comet could create the liquid primordial soup necessary for the formation of organic molecules. It is also possible that this primordial soup would allow the evolution of the simplest of life-forms, the virus.

Viruses, unlike meteor particles, would not burn up due to air friction upon entering the atmosphere. The virus cloud would slowly settle to the earth's surface. And where it settled, an epidemic would break out among the unprepared, unsuspecting population. There would be no immunity to the new virus from space.

The assignation of the heavens as an influence in the occurrence and spread of disease is actually centuries old. If the diseases-from-space hypothesis survives the passage of time, a cycle of thinking will have been completed.

The "flu," for instance, is a shortened form of "influenza," a term introduced in fifteenth-century Italy to describe an epidemic thought to be due to a malign "influence" of the heavens. The word "disaster," which aptly describes major epidemics, literally means "ill-starred," an evil influence of the stars.

Hidden astrology's biggest holiday is Christmas. Thanks to the Gospel according to Matthew, Christmas is the holiday of the year that has the strongest association with astronomy. Matthew probably no sooner crossed his last "t" and dotted his last "i" before someone started wondering what kind of star he was talking about.

Everyone wants to know about the "star of wonder, star of

light, star of royal beauty bright." Astronomers are usually very cooperative. They see the public's curiosity about the Star of Bethlehem as an opportunity to promote their favorite subject, which is astronomy. Christmas astronomy is annually promoted and sustained by greeting cards, songs, decorations, public lectures, newspaper and magazine articles, and star theaters.

Major planetariums since their inception have provided the public with numerous variations on the Star of Bethlehem theme. The reason for this is clear. Economics. Christmas shows draw the largest audiences of the year. And this is true year after year.

Astronomers take advantage of this seasonal enthusiasm to be in the spotlight. They use it to reveal what is known of the universe. They actually reveal little about the Star of Bethlehem because astronomers know very little about the nature of the Star of Bethlehem. They only know the "Star" of Bethlehem had nothing to do with astronomy. Furthermore, astronomers are secretly very glad of this.

The typical Christmas planetarium program or lecture is negative in content. It tells you what the star was *not*. The star was *not* a normal star, for instance. Normal stars are fixed in space as we see them and seldom change. The stars in our sky tonight are the same stars that were over Bethlehem two thousand years ago. Since the "Star" of Bethlehem attracted attention, it had to be something special. It could not have been a normal star. But what could this special thing have been?

The star was *not* a supernova because a supernova, the explosion of a star in its death throes, would have produced a new bright star in the night sky. It would have been observed by everyone, including Herod's stargazers. But according to Matthew's account, Herod's stargazers were unaware of the star that brought the Magi from the East to Jerusalem. The Magi add to the astronomical content of the story. They were dream interpreters and astrologers as well as experienced stargazer-priests of the ancient Babylonian religion of Zoroastrianism.

Herod's stargazers would also have been aware of the ominous presence of a comet, a snowball-like temporary visitor to the inner solar system. Other short-lived astronomical phenomena such as meteors, which are fragments of extraterrestrial debris burning up by friction as they plunge into the earth's atmosphere, are equally unsatisfactory, simply because of their transient nature. The Magi followed the "star," so it had to have been visible over an extended period of time.

Johannes Kepler, the seventeenth-century German astronomer who discovered the elliptical nature of the orbits of the members of the solar system, proposed that the star was actually a triple conjunction between Jupiter and Saturn that was followed by the appearance of Mars to produce a spectacular three-plant triangle in the zodiac constellation Pisces the Fish. Triple conjunctions of planets, the passing of one planet by another three times due to its orbital motion and its apparent retrograde motion near the time of opposition with the sun, are rare events. The last triple conjunction between Jupiter and Saturn occurred in 1981. The next one will take place in 2238. The triple conjunction has been a popular explanation for the star. This has been especially true in planetarium presentations because the triple conjunction can be visually demonstrated in a planetarium. But like all other astronomical explanations, it is hard to see how it could have led the Magi to a particular dwelling.

So what *was* the star? It was nothing that can be identified as astronomical. This seems to be a safe statement. But look at all the astronomy you were introduced to in bringing you to this point. Q.E.D.

Astronomers are relieved that the Star of Bethlehem does not have an apparent astronomical explanation. The use of the triple conjunction and the Magi in the telling of the "astronomy" behind the Christmas story makes astronomers somewhat uncomfortable, since it lends a certain credibility to the claims of astrology.

The basic claim of astrology is that celestial events can foretell terrestrial events. If the Star of Bethlehem were indeed a celestial event, it would provide astrology with the greatest big-name endorsement it could ever hope to receive. Astronomers would rather the possible explanation for the Star of Bethlehem be found in terms of a literary device or a miracle than some regular event in the sky.

Perhaps Matthew, the only Gospel writer to mention the star, used it as a literary device in order to reinforce his story about the birth of Jesus. Perhaps the star was a miracle, an event that defies the known laws of nature. Perhaps the star that the Magi followed to Jerusalem was only seen by them. If the nature of the Star of Bethlehem was known for sure, history's most famous UFO case would be solved.

There is another day of the year containing an element of hidden astrology. Groundhog day is the day of the halfway point between the winter solstice and vernal equinox. On this day the groundhog or woodchuck is given meteorological predictive powers, according to the traditional superstition. If a groundhog leaves its burrow on the appointed day in the beginning of February and sees its shadow, it will return to its burrow and winterlike weather will continue for six more weeks. A sign from the heavens is used to predict the future. This is astrology. This is crystal-ball astronomy.

A belief in astrological meteorology was so strong in the past that it influenced the naming of the dark areas visible on the face of the moon. The dark areas seen on the lunar surface that form the "man in the moon" were called *maria* (the plural of *mare*) or seas by Galileo after observing them with his telescope in the early seventeenth century (see pp. 72, 168–169).

The maria are not seas of liquid water, as Galileo thought. They are frozen seas of once flowing hot lava that rose through fractures in the lunar crust. The different maria were named by seventeenth-century lunar mapmakers according to the astrological superstition that the first quarter moon promoted good weather, while the last quarter moon promoted bad weather. The western hemisphere of the moon, which is visible during the first quarter waxing phase, contains the seas of Serenity, Tranquility, and Fertility. These are names associated with good weather. The eastern hemisphere of the moon, which is visible during the waning phase, contains the seas of Showers and Clouds and the Ocean of Storms. These are names associated with bad weather.

A belief in an astrological calendar was the key that opened the doors of isolationist China to the West in the seventeenth century. A solar eclipse took place on June 21, 1629, which served as the object of a competition, and the final score was Jesuit missionary astronomers 1, Chinese astronomers 0.

The Chinese lunisolar calendar is among the most ancient in recorded history. It goes back more than four thousand years. But its development was retarded, and in fact diminished, by the removal of Chinese astronomy from the mainstream of Chinese thought and knowledge. This is why the Jesuit missionaries won and why the competition took place in the first place.

The emperor of China was acknowledged as the "Son of Heaven." He ruled as part of the natural celestial order. If he was not a good ruler, then celestial events would warn against the transgressions of the celestial will. The duties of the Grand Astrologer of the Chinese Astronomical Bureau were concerned with anticipating important celestial events.

The Grand Astrologer's most important duty was that of producing an annual astrological calendar. This calendar was important because it functioned as a form of portent astrology. At the beginning of each year the emperor was presented with a calendar listing the astrological prognoses for the coming year. It provided the propitious hour and day for the performance of state duties and actions. If an activity were performed at the improper time, it was believed that it would not succeed. Worse yet, bad luck would follow.

The astronomical knowledge needed for the construction of

the emperor's calendar was officially restricted to a small group of astronomer-astrologers. The proficiency needed for the construction of this most important state document was a top secret skill limited to those with a "need-to-know" clearance since it was thought to contain information that could impact upon national security.

It was believed the calendar could be a dangerous weapon to be used against the emperor if it fell into the wrong hands. It was feared that if astronomical knowledge and unauthorized private calendars got into the hands of the common people, political unity and the imperial authority could be jeopardized. The knowledge of calendar-making was therefore controlled and confined to members of the imperial court.

The prognostications of the calendar were empirically derived. "What" was known, but not "why." Over a long period of time astronomical and meteorological phenomena had become correlated to happenings in human society relevant to the well being of the emperor. Over the same long period of time astronomical and meteorological phenomena had also become viewed as good and evil omens. But the derivation of these omens were without a basis in logic. Chinese calendar-making and astrology did not lead to an objective study of the heavens. It did not investigate new approaches. It did not develop a theoretical base. It focused on anomalies rather than regularities. And as a consequence, in time, the calendar-making rules were lost.

The Chinese calendar was periodically checked by comparing the prediction of an event, such as a solar eclipse, with the actual observation of the event. Solar eclipses were particularly important because they were considered evil omens. The predictions had to be correct so countermeasures could be taken to avert any impending disaster. Revisions were then instituted whenever a discrepancy became noticeable.

On the day preceding the solar eclipse of June 21, 1629, the Chinese astronomer-astrologers and the Jesuit missionary astronomers submitted their predictions in writing for the time and duration of the next day's solar eclipse. The Chinese astronomer-astrologers predicted a two-hour total eclipse starting at 10:30 A.M. The total solar eclipse lasted two minutes starting at 11:30 A.M., as predicted by the Jesuit missionaries.

The outcome of this contest influenced the course of historical events in China. The Chinese needed help and they knew it. Furthermore, they knew where to find the help they needed. The task of reforming the Chinese calendar was given to the Jesuit missionaries and their students. On February 28, 1634, the first calendar based on European methodology was presented to the emperor. China had opened to Western influence.

For a long time people have looked to the sky as they might look into a crystal ball for the answers about tomorrow. The supposed predictive powers and influences of the sky have been strongly promoted and believed in the past. It should not be surprising to discover they have quietly infiltrated into other areas commonly believed to have predictive powers.

Consider, for instance, the long tradition of wearing a gold wedding ring on a certain finger, the "ring finger," of the left hand. Could astrology possibly have had an influence in establishing this tradition? The answer is yes. The sky has influenced even the traditions of palmistry.

Palmistry is a method of folklore prediction that uses the form and lines of the hand for its "readings" of the past and the future. Forklore tradition also has linked the parts of the hand, as well as the lines on the hand, to the sun, moon, planets, and signs of the zodiac.

During medieval times, palmists identified the left-hand ring finger with the beneficent sun. The sun was the ruler of spiritual life, and gold was believed to be the metal representative of the sun. It was also believed that this finger contained an artery that ran directly to the heart. And the heart was considered the center of emotional life. So, it was believed that giving and placing a gold ring on the ring finger of another person established a beneficent mystical connection between the emotional and spirtual life of the wearer and giver. There it is, the sky, astrology, and the wedding ring. It sounds a little farfetched, but that is because the origin of the practice has been forgotten.

Rings are not the only things worn that can have subtle or hidden astrological connections. Amulets and charms are also predictive devices that can have connections with the sky. Amulets and charms supposedly ensure (predict) future safety from the influence of evil by their ability to ward off evil. Or, in another mode, they ensure (predict) future "good luck" for the wearer by attracting good fortune. Many of these predictive devices are not recognized as such today because they have been incorporated into the design of jewelry. We see them as jewelry designs and not as amulets and charms. With time their original use has been forgotten.

The most popular amuletic symbol with an obvious astronomical connection is the five-pointed star. It is seen prominently displayed on flags, official seals, and military uniforms. The five-pointed star can be found almost anywhere. The use of the star symbol is an unconscious carryover from the time when its appearance supposedly provided a protective power from the goddess Venus. Zodiac jewelry, displaying the signs of the zodiac, served the same purpose. It called forth the supposed powers and influences of the signs to ensure (predict) a good or safe future.

Alchemy, a form of mystical chemistry, also had a predictive tradition with sky associations. The seven alchemical metals—gold, silver, mercury, copper, iron, tin, and lead—were believed to reflect the nature and influence of the sun, moon, and five naked-eye planets. The balance between these elements inside the human body was believed to determine the temperament, personality, and appearance of individuals. It supposedly produced seven different types of people representing the individual influence of each of the seven celestial bodies.

Crystal ball astronomy, in the form of astrology, is all around us even today. You can thank your lucky stars if you have been able to escape its influence.

7

Astronomy Since 1610

In the early seventeenth century something happened in the Western Hemisphere. And as a result, astronomy in the Western Hemisphere was divorced from astrology. The something that happened was the invention of the telescope.

With the invention of the telescope, astronomy in the West entered Lockyer's third stage. The sole purpose of Lockyer's third stage was the observation of celestial objects for the acquisition of knowledge without regard for its potential or eventual use. Astronomy became research-oriented, and the emphasis on utilization diminished. Astronomy came to be supported in the West by government agencies, educational institutions, and private resources, and no longer had to depend on astrology.

Galileo Galilei (1564–1642) was the first important astronomer in this stage of development in the history of Western astronomy. He was the first to scan what had been the naked-eye universe with the added sight of the telescope. In 1610, Galileo was what would today be described as a professor of physics at the University of Padua in Italy. In that year he received a report about a Dutch lens maker who was able to make distant objects appear closer by the use of two glass lenses. Galileo quickly built his own superior version of this "optik tube."

While others used the "optik tube" to bring earthly objects nearer, Galileo turned his new telescope to the heavens. He was able to see what had never been seen before. Galileo's telescopic discoveries had a tremendous effect upon the thinking of his time. They helped to establish the viability of the new, sun-centered Copernican model of the universe and call into question the long-established, earth-centered Aristotelian model.

On January 7, 1610, Jupiter appeared to all the world, except for one man, as a magnificent shimmering starlike object in the constellation Taurus the Bull. Galileo looked at Jupiter through the telescope and saw not a star, but a disk. This immediately distinguished Jupiter from the more distant stars that still today appear as points of light in even the most powerful telescopes.

But what attracted Galileo's attention most were the four "starlets" that appeared from night to night to move from one side of Jupiter to the other. The four "starlets" were moons in orbit around Jupiter. Galileo had discovered the four largest natural satellites of Jupiter.

Galileo's small ten-power telescope had discovered in Jupiter a miniature replica of the Copernican model of the solar system. Jupiter was a system in which objects of lesser mass were in orbit about a more massive object. The Jovian system had the added feature of having the inner moons moving faster than the outer orbiting moons, just as the planets moved in the solar system.

The moons of Jupiter were a significant discovery. Until this time, it had been argued that the earth was the center of the universe, since all observed motions of the heavenly bodies were centered on the earth. Jupiter showed this precept was not totally accurate.

But more significant was the discovery that the moons moved along with Jupiter, as Jupiter itself orbited the sun. The moons of Jupiter helped to silence those who thought it impossible for the earth's moon to go around the earth and simultaneously keep up with the earth while it went around the sun. Jupiter proved it could be done.

Galileo used his further telescopic observations as additional proof to show that the two-thousand-year-old Aristotelian earth-centered view of the universe was seriously flawed. Galileo looked at the moon and saw earthlike features. He saw mountains, valleys, craters, and large smooth areas he mistakenly interpreted as bodies of water. He called them "maria," meaning seas. This was a significant discovery, because prior to this observation, the moon was considered to be a perfectly smooth sphere, totally unlike the earth.

Galileo turned his telescope on the stars and discovered there were stars in the night sky that had not, and that in fact could not, be seen with the naked eye. He discovered that the glow known as the Milky Way was "a mass of innumerable stars planted together in clusters." This discovery provided important information about the potential distances of the stars and made Copernicus's assertions about the great size of the universe more plausible.

The most important observational discovery made by Galileo concerned the phases of Venus. The Copernican theory predicted Venus would exhibit the same waxing and waning phases as the moon if Venus traveled around the sun. The older, established tradition of Aristotle predicted Venus would only show crescent phases to an earth observer because Venus, in the Aristotelian system, was always situated between the earth and the sun. The phases of Venus provided a critical test between the Copernican and Aristotelian models.

Galileo's telescope revealed Venus in the same sequence of waxing and waning phases as the moon. This discovery proved conclusively that the Copernican model of the solar system was superior to the Aristotelian model, but it would take a lot more time to actually prove the earth both rotated and revolved.

Galileo's celebrated problems with the Catholic church had little to do with whether the earth or the sun was the center of the universe. His problems had everything to do with authority. Tradition and established authority are always adversaries of change. This was especially true in the time of the rise of Protestantism.

Galileo raised the question of how a decision was to be made on the construction of the heavens. Was the system of the world to be decided by the authority of reasoning from observations alone, as Galileo was advocating? Or was the system of the world to be determined by the authority of tradition and established procedures within a conservative religious framework, as had been the case in the past? It is too easy to pass judgment by contemporary standards and practices.

The times, the man, and the issues were complex. This is why so many myths have surrounded Galileo, including the most popular one that he muttered, *"Eppur si muove!"* ("And yet it moves!") following his forced recantation of his Copernican views. There is no basis for believing that he did this, nor that the Inquisition would have permitted him this final act of defiance.

Galileo had no conclusive proof that the earth rotated or revolved. His Copernican stance was based mostly on faith. Galileo's 1633 heresy trial did not involve the question of the movement of the earth. It involved the movement of opinion as to how truth was to be discovered, but it created the myth of an inevitable battle between religion and science. Galileo, a religious man, was not fighting religion as such: He was fighting an established conservative authority. There is a difference. He lost his battle, but he contributed to the winning of the war because he had the unmitigating gall to be correct.

Galileo's telescopic discoveries provided the world with a "golden age" of space exploration that has only been rivaled in the twentieth century. Galileo was the middle of a triumvirate of three great men who completely changed the concept of what we see in the night sky. The other two men were Nicholas Copernicus and Isaac Newton. Copernicus developed an idea and model, Galileo provided the supporting observations, and Newton constructed the synthesis that tied it all together. It is historically interesting to note that Copernicus died in the year Galileo was born, and Galileo died in the year Newton was born.

The historical line from Galileo to Newton is not direct. It passes through the contributions of a Danish nobleman, Tycho Brahe, and a German mathematician, Johannes Kepler.

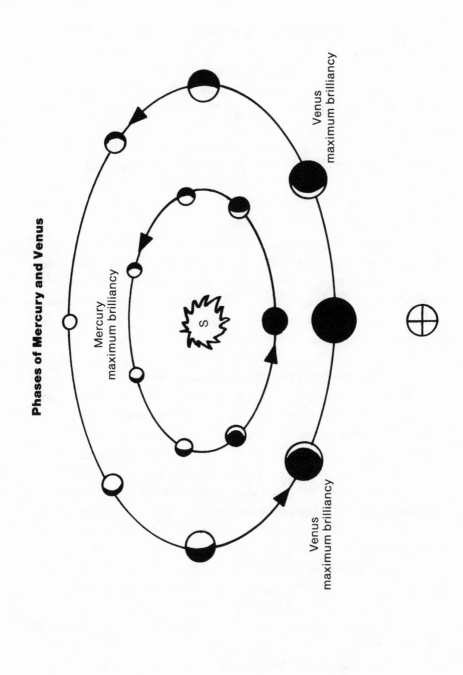

Phases of Mercury and Venus

Mercury
maximum brilliancy

Venus
maximum brilliancy

Venus
maximum brilliancy

Tycho Brahe (1546–1601) is recognized as the greatest naked-eye astronomer to have ever lived. He died a decade before the invention of the telescope. He was also the most flamboyant astronomer to have ever lived. From the events following his birth to the events preceding his death, nothing that he did and nothing that happened to him was ordinary.

In the Brahe family's version of "Rumpelstiltskin," Otto and Beate Brahe promised if they produced a son the child would be placed in the care of a fraternal uncle to be raised as the uncle's own son. When Uncle Jorgen showed up to collect on the promise, Tycho's parents had second thoughts.

If it hadn't been for an earlier tragedy, the situation could have been resolved in a very civilized manner. Tycho had a still-born twin brother. If the brother had lived, there would have been enough Brahe sons to go around, and everyone would have been satisfied. As it turned out, Uncle Jorgen believed a promise was a promise, so he kidnapped young Tycho from his crib and retreated behind the locked gates of his castle. Tycho's father threatened to murder his brother if the child was not returned.

Mother Nature resolved the dilemma. Beate Brahe became pregnant again soon after Tycho's birth and delivered another son. The crisis was resolved; there were now enough Brahe boys to go around. Eventually there were more than enough to go around. The second Brahe son was followed by three more.

It wasn't until Tycho was twenty-six years old that he decided to devote his life's work to observing and measuring the sky as accurately as had ever been done before. The contributing event was the appearance of a new star, or "nova," in the constellation Cassiopeia the Queen. The new star was so bright it was possible for it to be seen in the daytime sky. After eighteen months, the star disappeared from sight. We know today the star only appeared to be new; it was actually an old star exploding in the supernova throes of dying.

Tycho realized that the controversy between the Artistotelian model, Copernican model, and eventually his own compromise Tychonic model could not be adequately resolved with the continued use of observations dating from the second century B.C. by the ancient Greek astronomer Hipparchus. A new set of accurate observations was needed. Toward this end, Tycho was able to build a royally supported, palatial observatory on the island of Hveen off the coast of Denmark. Uraniborg, the Castle of the Heavens, was the most famous observatory of all times in terms of grandeur. Among its distinguishing features were gold decorations, talking statues, and a clown dwarf, Jepp.

Tycho Brahe's ability to build naked-eye observing instruments and his twenty years of continuous observation produced a catalogue of reliable positional data for 777 stars, the sun, moon, and planets. The accuracy of Tycho's catalogue was challenged only by the invention and use of the telescope for positional astronomy.

It was drink and good manners that sent Tycho to his grave. At the beginning of the seventeenth century, it was considered ill-mannered to leave the table of one's host to relieve the pressure on one's bladder. Tycho put his courtesy, his appetite, and his love of drink before his comfort at a big dinner one night, and as a result died within a few days.

The value of an astronomer's service can often be measured in the contributions he provides to the work of those who follow. Tycho had an effect on the history of astronomy beyond the production of his catalogue. Tycho's extremely reliable data concerning Mars eventually lead to Johannes Kepler's solution to the problem of how the planets move around the sun.

Johannes Kepler, a contemporary to Galileo, used Tycho's precise observations to overturn the two-thousand-year-old tradition of describing the planets as moving in uniform circular motion. Kepler used Tycho's observations of the mystifying movement of Mars through the background stars to discover the elliptical shape of the orbit of Mars. He also discovered how Mars moved in its elliptical orbit. The description of the orbit and motion of

Mars was the description of the orbits and motions of all the planets. Kepler produced the first set of mathematical statements or laws to describe the motions of celestial objects.

One of Kepler's laws states that the squares of the periods of revolution of the planets are to one another as the cubes of their mean distances from the sun. Only the Pythagorian belief that "all things are number" would lead to a discovery of this celestial harmony. Kepler believed that God was a mathematician and that the secrets of the universe could be discovered in mathematics.

Kepler, for instance, believed that there were only six planets arranged as they were in relation to one another because there were only five different perfect three-dimensional geometric solids. The connection is not obvious, but it determined the course of Kepler's work in astronomy. It was a belief that dominated his career.

A perfect geometric solid is one in which all of the faces are identical. The cube is the best known example. The symmetry of the five perfect solids is such that each can be put into a sphere of the proper radius so all of the corners touch the sphere. Five perfect, geometrical solids within spheres were seen by Kepler to determine the relative distances of the six known planets.

The sphere of Saturn had a radius equal to the distance of Saturn from the sun. Within Saturn's sphere, Kepler placed the biggest cube possible. Within this cube, he placed the biggest sphere possible. This was Jupiter's sphere. The numerical ratio of the radii of these two spheres was equal to the ratio of the approximate distances of Saturn and Jupiter from the sun.

Jupiter's sphere had the largest tetrahedron possible (four equal-sided pyramid) placed within it. To this, Kepler added the largest sphere that would fit within the tetrahedron. This was Mars's sphere. The correct radii ratio continued.

Kepler obtained equal success for the earth by using a solid of eight equilateral triangles called an octahedron. For Venus he used the twelve pentagons that form a dodecahedron, and for

Mercury, the twenty equilateral traingles of an icosahedron.

Kepler could not believe this arrangement of six planets separated by five perfect geometric solids was the product of chance. To him, it revealed the workings of God's mind. God was a mathematician and since God could only create a perfect universe, and since there were only five perfect geometrical solids, he placed them between a necessary six planetary orbits to create a perfect system. God mercifully allowed Kepler to die before the seventh planet was discovered.

Isaac Newton, late in the seventeenth century, produced a magnificent synthesis of the observed universe with his mathematical theory of universal gravitation. It is a well-known story. An apple fell on Newton's head, and he discovered gravity.

Newton, who had absolutely no sense of humor, would have found this statement side-splittingly hilarious. An apple never fell on Newton's head, and the discovery of "gravity" wasn't that simple. And yet, this "Newton and the Apple" myth is one of the best known episodes in the life of the seventeenth-century English natural philosopher.

This myth, like all myths, has some truth buried within it. Newton, by his own admission, did see an apple fall to the ground, and he did think about gravity as a result, and he did wonder if gravity held the moon in its orbit around the earth. But he did not discover gravity. He only changed the meaning of the word.

"Gravity" was a word in use long before the apple incident, but it meant simply "an affection or desire." Apples fell to the ground because they had an affection for the earth, or a desire to return to the earth out of which they were composed. There was a place for everything and everything wanted to go back where it came from. Newton changed the word "gravity" to mean a force of attraction between two masses.

Newton proposed that what was called gravity was an attractive force that the earth exerted on the apple to make it fall toward the earth. He proposed that the earth also exerted an attractive force on the moon to make it "fall" toward the earth.

Newton knew the moon needed an attractive force acting on it to keep it in orbit around the earth. Without a force pulling it toward the earth, the moon would go flying off into space on a straight line path. A force was needed to keep the moon falling toward the earth at the same rate the curved earth was falling beneath the moon. In this situation, the distance between the falling moon and falling earth stayed the same. The result was the moon in orbit around the earth.

Newton's problem was in calculating a value for the attractive force of gravity at the distance of the moon. He already knew the value of the force required to keep the moon in orbit. The question was whether it was gravity that supplied the force needed. This was not an easy task for a twenty-three-year-old who had just finished his undergraduate studies. In attempting to solve the problem, Newton had to invent calculus. There have been multitudes of college students who have never forgiven him for this indiscretion.

Newton's final answer was initially disappointing. In his mind, the calculated force of gravity at the moon was close to the force needed, but "no cigar." So he dropped the problem for another twenty years until Edmund Halley asked him what he thought kept the moon in its orbit. The problem with his earlier calculation was in the inaccurate value he used for the size of the earth. The value was corrected, however, by the time of Edmund Halley's inquiry.

With Halley's support and financial sponsorship, Newton revealed to the world the inner thoughts of his mind concerning the universe and how it operated. This revelation produced the most influential book ever written. Since its publication it has dominated the way the universe is perceived throughout the entire learned world. The book was titled *Mathematical Principles of Natural Philosophy*. It is best known by its shortened title, the *Principia*.

Newton used the *Principia* to paint a masterful picture of the universe in mathematical symbols. His synthesizing system of thought involving a universal gravitational force provided such a beautiful vision of the universe that the world has never fully recovered. The scientific revolution that was begun by Copernicus, Kepler, and Galileo was indisputedly achieved by Newton.

While the *Principia* is virtually unread today, the physical laws and scientific principles contained within its pages can be found in use in any science classroom or scientific laboratory anywhere in the world.

Few people have never been exposed to Newton's laws as handed down in the *Principia*. His law that "a body at rest will remain at rest, and a body in motion will remain in motion unless acted upon by some outside force" is recited confidently by every physics student in spite of their lifetime experiences to the contrary.

His law expressed as $F = MA$ is so well recognized as "science" that it is placed on blackboards by cartoonists to create the illusion of a science classroom or laboratory. And the phrase "to every action there is an equal and opposite reaction" is recognized as a Newtonian decree by many more people than understand it.

The *Principia's* contribution to the conduct of the scientific enterprise, while as influential as its physical laws, is not as well recognized. It was Newton who proposed that science should proceed on the principle that the same effects arise from the same causes. This principle is the foundation of experiments performed with laboratory subjects to better treat and understand human physical and psychological maladies.

His belief that nature is basically simple, and that any description of the workings of nature should contain no more hypotheses than necessary, is unquestioned in science. The same is true of his proposition that hypotheses accepted on experience should only be thought of as very nearly true. Because of Newton's influence, science as practiced throughout the world knows of no absolutes.

The most revolutionary of Newton's contributions to science, and the unifying glue of his world picture, was his proposal of a law of universal gravitation. It involved the interaction of bodies in space through an attractive force that was directly proportional to the product of the masses of the bodies involved and inversely proportional to the square of the distance between the bodies.

The irony in the influence of Newton's view of a universe held together by the attractive forces of gravity is that there is no such thing as the gravity force it proposes and uses to explain the universe. Einstein replaced gravity with a curved space universe, but Newton's laws of universal gravitation are still taught and obeyed. Apples are still thought to fall to the ground because of gravity.

There is another irony in Newton's life and it involves his godlike status. The following clues should bring a historical figure to mind:

CLUE ONE: This person's birth is remembered near the end of December at the peak of the holiday season. Many people attend church services on this day. This person died near the time of the Passover.

CLUE TWO: Few people who lived in the area were aware of the birth at the time it took place. Those present at the birth had reason to believe this child was a child with a destiny.

CLUE THREE: This person's father was not physically present at the time of birth. He was supposed to be in heaven.

CLUE FOUR: There was a bright light in the sky at the time of this person's birth.

CLUE FIVE: Farm animals were nearby, since the birth took place in a simple farm dwelling. The nearby town was little.

CLUE SIX: This person's life was in danger at the time of birth and for a period following the event of the birth. It was a time of civil unrest, and soldiers were in the area.

CLUE SEVEN: The king in power at the time of this person's birth would die a few years later.

CLUE EIGHT: This person was different from other people. This person never married.

CLUE NINE: This person's main concern throughout life was religion. The Old Testament was this person's favorite book.

CLUE TEN: This person's name is recognized throughout the world as an authority figure. It is not unusual for people to memorize the laws this person formulated to replace the laws in use at the time of this person's birth.

The person described is Isaac Newton. Newton was born between 1:00 and 2:00 A.M. on Christmas day, 1642. He died between the same hours on March 20, 1727, near the time of the Jewish Passover celebration. People in the area were unaware of his birth because Newton was born six weeks premature.

Newton's father, who was also named Isaac, was not present at the birth because he had died two months earlier. It was for this reason that it was believed that Newton was a child with a destiny. According to a local Lincolnshire belief, this was true of any male child born after his father's death.

The bright light that was in the sky at the time of Newton's birth was a full moon. The Newton house was the manor house of the little town of Colsterworth, but it was still a simple farm dwelling. The barn was attached to the back of the house, so Newton's birth was probably heralded by the moo-ing of cows and the baa-ing of sheep.

Newton's life was in danger at the time of the birth and for weeks after because he was born six weeks premature. Fear for the survival of the weak and frail child was responsible for his speedy baptism in the parish church at Colsterworth on New Year's Day 1643. Newton survived for another eighty-five years.

Although a hypochondriac, Newton enjoyed good health, keeping his hair, teeth, and eyesight. It was a stone in his bladder that caused his death.

The time of Newton's birth and early years was a time of political unrest. England in 1642 was experiencing a civil war.

Oliver Cromwell's forces fought the forces of Charles I only ten miles from the Newton home when Newton was five months old. Charles I was beheaded when Newton was seven years old.

Isaac Newton certainly qualified as being different from other people. He was the premier academic genius, but he was socially retarded, especially with respect to women. The only women in his life were his mother in his early years and a favorite niece in his old age. It has been suggested that for some strange reason Newton had a messiah complex.

Surprisingly, Newton's main concern in life was religion, not science. He was especially interested in Old Testament chronology and the prophecies of Daniel. Newton actually produced a much greater volume of works about religion than about physics and astronomy. But it is in physics and astronomy that Newton's genius is recognized.

Edmund Halley demonstrated the validity of Newton's laws by successfully predicting the return of a spectacular comet with a period of seventy-six years. He hoped that "candid posterity will not refuse to acknowledge that this was first discovered by an Englishman," for he knew he would not be alive to say, "I told you so." Candid posterity did not forget. The returned comet was given Halley's name.

Unfortunately, this is mostly all that "candid posterity" has remembered about Halley. He deserves better. If it were not for Halley, there is a chance that Isaac Newton would have remained a minor figure in the history of science. It was Halley who uncovered Newton's genius and revealed it to the rest of the world. It was Halley who provided time and money to oversee and publish Newton's *Principia*.

The discovery of Newton was not Halley's only accomplishment. His career reflected the breadth of his interests. Halley was an accomplished scholar in such widely diverse studies as archaeology and classical literature, as well as most branches of the physical sciences.

When Halley was appointed the Astronomer Royal of England at the age of sixty-five, he set about attempting to solve England's most pressing maritime problem. This was finding the position of longitude at sea. The problem was a commercial and military concern to all seafaring nations. The solution to the problem required a method of accurately determining time from the deck of a moving, pitching ship at sea. Halley, as well as Newton, rejected the idea of a mechanical clock as a possible solution. They favored, instead, an astronomical solution in which time could be accurately determined by observing the heavens.

Halley's proposed astronomical solution required an exact knowledge of the moon's motion through the background stars over one complete cycle. Halley called the bluff of Father Time when, at age 65, he set out to completely observe the 18.6-year lunar cycle. He won the bet, having observed every clear night until the 18.6-year cycle was completed. It was then that his health began to fail. Halley's lunar observations provided British ships with a practical method for estimating, though not accurately determining, their longitude when they were out of sight of land.

But it is from the comet bearing his name that Halley is remembered. Conventional wisdom at the beginning of the eighteenth century described comets as one-time-only celestial events. The question of possible returns of the same comet arose when Halley discovered that the comets of 1531, 1607, and 1682 seemed to have similar elliptical orbits. These comets had also appeared after almost equal intervals of time. Halley predicted that a great comet would appear in 1758 and that it would represent the return of a past comet following a period of seventy-six years. The returning comet was first sighted on Christmas night 1758 by Johann George Palitzsch, a German amateur astronomer.

An age can accept only one genius, and because of this, the genius of Halley has been forever lost in Newton's shadow. When Newton died, he was buried in a place of high honor in Westminster Abbey. Thousands of tourists view his resting place every day.

Halley's body lies in a small church cemetery one mile from the Greenwich Observatory. It is seen each day only by the regular group of commuters that wait for their morning bus within sight of his uncelebrated resting place.

While astronomy was developing along mathematical lines, and telescopes of increasing size were being built in the Western Hemisphere to extend the visible horizon of the universe, almost nothing was happening in the Eastern Hemisphere. The exception was in India, where an early eighteenth-century Mogul emperor, the Maharajah Jai Singh, was concerned with building a large series of sundials. The Jantar Mantar Observatory in Delhi was the first of five to be built at places where the Maharajah served as governor or ruler. Others were located at Jaipur, Banares, Ujain, and Mathura. Jai Singh was interested in astronomy in a big way.

The largest instrument at Jantar Mantar consists of a huge triangular masonry gnomon measuring sixty feet in height with a scaled quadrant on both sides. This Samrat Yantra, meaning "supreme instrument," is a garden design sundial built on a large scale. But it simply measures the local apparent solar time. Other instruments at Jantar Mantar can be extended to determinations of the changing position of the sun with respect to the background stars and horizon. These instruments also can determine the celestial positions of other bright objects, such as the moon, bright stars, and planets.

Without a doubt the development of the sundial reached its height in both size and sophistication in early eighteenth-century India. But what was Jai Singh's purpose in building his oversized sundials? It was not to display his creativity. His sundials are variations of previously proven designs. The novelty of Jantar Mantar is in its proportions, not its originality.

The size factor, however, could have resulted from a desire to reduce the inherent errors in such instruments as much as possible. It could also have arisen from his need to display his wealth and power. Maharajahs have never been known for their avoidance of excesses. The only astronomical purpose for building his sundials seems to lie in the possible construction of a new set of astronomical tables that could have been used for calendrical purposes.

Maharajah Jai Singh was a contemporary of Isaac Newton and Edmund Halley. He could not have been ignorant of British accomplishments in astronomy. It seems reasonable to assume that Jai Singh's interest in astronomy would have been fully exploited by the British whose influence in India was expanding at the time. This being true, it is difficult to explain why he ignored the use of the telescope, which had been invented a century before. But then India has always marched to its own tune.

Time has all but erased from historical memory the reading of Jai Singh's thoughts. It has treated his sundial observatory with the same disdain. The sundials can no longer be read as easily as in the past. The gradations on the lime plaster dials have all but disappeared, leaving behind a Mogul Indian monument to and of the passage of time.

Meanwhile, in the West, Lockyer's third stage continued to advance full force. On March 13, 1781, William Herschel (1738–1822), a professional musician who was an amateur astronomer, made a startling discovery. He thought he had discovered a comet in the constellation Gemini the Twins, but, in fact, he had made the first discovery of a planet in historic times. The discovery was made with the assistance of a telescope. The planet was called "Georgium Sidus" (King George III's star) by Herschel, "Herschel" by others, and finally "Uranus."

Herschel's interest in astronomy was developed by his father. When Herschel was a child his father taught him to identify the nighttime stars and constellations. Herschel's obsession with astronomy, however, came by way of his professional involvement in music. The theory of musical harmony led Herschel to an interest in mathematics, and the study of mathematics led him to an interest in optics and astronomy. The study of astronomy produced a desire to see the universe through a telescope. Since good telescopes were too expensive to purchase, Herschel decided to make them. By the end of his life, he had made more than four hundred telescope mirrors. His largest mirror was forty-eight inches in diameter.

Following the discovery of Uranus, George III, King of England, elevated Herschel to professional status by appointing him a Royal Astronomer and providing him with the financial support to build an observatory near the royal residence at Windsor Castle. Herschel went on to become the most productive astronomer of his time. This was because he continued in the tradition of an amateur astronomer. He continued to be a stargazer.

Professional astronomers had become problem solvers. They had begun to use the sky to gather data for a solution to a particular piece of the puzzle involving the nature of the universe. This problem solving research tradition continues today.

Amateur astronomers are stargazers. They are the true disciples of Lockyer's third stage in the development of astronomy. They observe the sky to see, or record, what is there without concern for doing something with their observations. Herschel went on to observe and catalogue the universe visible within giant reflecting telescopes of his own design and construction. He went on to produce extensive sky catalogues from his systematic observations of the night sky, while at the same time using these observations to produce hypotheses to answer such questions as the shape of the universe.

Uranus proved to be a problem for the astronomers who followed Herschel. When astronomers attempted to calculate an orbit for Uranus that would allow them to accurately predict its future positions among the stars, they were not successful. Many theories were proposed to explain the discrepant behavior of the planet. One theory proposed that an as yet undiscovered and more distant planet was gravitationally influencing the behavior of Uranus. The problem of determining the position of this unknown planet, if it existed, was considered by astronomers to be insurmountable.

The problem was solved with the discovery of Neptune in 1846.

The discovery of the eighth planet of the solar system is credited to the independent achievements of the English and French mathematicans John Couch Adams (1819–1892) and Urbain Leverrier (1811–1877). Unlike Herschel, they discovered Neptune without a telescope. They used paper and pencils.

Adams and Leverrier independently derived the positional coordinates for the disturbing force on Uranus from a purely mathematical treatment of the problem. But they both had a difficult time finding an astronomer to actually look for the planet. The astronomers had little faith in what they considered mathematical exercises based upon inspired guesswork. The Berlin Observatory eventually honored Leverrier's request and found the planet on the first night of searching. This incident marked the beginning of theoretical mathematical astronomy. Leverrier had his revenge on the astronomers who would not look for the planet. (He supposedly was never curious enough to look at Neptune in a telescope.)

Albert Einstein introduced the general theory of relativity in 1916, claiming that space is curved and that gravity is an unnecessary fictitious force resulting from the curvature of space. This has been one of the best kept secrets in the world. This secret is especially kept from beginning physics students. If they knew that gravity was just a figment of Isaac Newton's imagination, they would lose the incentive to mentally master Newton's laws.

Newton's law of universal gravitation states that all bodies in the universe exert a force of attraction on all other bodies in the universe. Logic dictates that these same bodies could just as well have exerted forces of repulsion on each other, or, as may seem more logical, they could have just ignored each other.

But no, Newton convinced the world that they exerted forces of attraction on each other. He further convinced the world that the magnitude of this attraction was directly proportional to the product of the masses of the involved bodies, and inversely proportional to the square of the distances between them.

Newton's law of universal gravitation was questioned at the time of its proposal. His colleagues may have thought it a bit presumptuous of Newton to attempt to tell them about the universe, when he had hardly been more than fifty miles from his birth place at anytime in his life.

Newton was asked a simple question. How can two bodies exert a force on each other without touching directly or through some intermediary? The question was simple, but it was also an embarrassment. Newton replied by stating, "I make no hypothesis." The fact is, he really didn't know the answer, but he still thought gravity was a good idea, or at least better than any other idea proposed.

Einstein proposed that the behavior attributed to gravity forces was really due to distortions of space. Large masses like the earth do not exert gravity forces; they distort space within their vicinity.

Bodies moving in distorted space follow paths requiring the least energy. Water flows downhill because that is the path of least energy for the earth's surface. It is easier for water to flow downhill than uphill. Apples fall to the ground because that is the path of least energy for the space surrounding earth.

Einstein's hypothesis was tested during a 1919 total solar eclipse of the sun. It was a critical test involving a star that was known to be just behind the solar disk at the time of a total solar eclipse. This star could only be seen at the time of totality if its light did not follow a straight line path. The light would have to curve around the sun.

Gravity could not make this star visible during the eclipse because light does not have mass. Gravity does not affect the motion of a massless light beam. If the light from this star was seen during totality, then the starlight would have traveled a curved path around the sun on its way to earth because of the distortion of space close to the solar surface. If the star was not seen during totality, then space was not curved. The star *was* seen. Space *is* curved.

Einstein's theories of relativity extended the work of Isaac Newton to situations, that in Newton's time, were beyond comprehension. The results of Einstein's mathematical investigations are often beyond today's common sense. We exist in Einstein's relativity universe but we are most comfortable in Newton's description of it.

Other twentieth-century investigations have also extended our description of the universe to the limits of what seems like common sense. M31 is an example. Of all the celestial objects visible to the naked eye, M31, a faint glow of light in the Andromeda constellation, has produced the most significant change in our description of the universe.

M31 derives its notation from its being the thirty-first object on a list compiled by the late eighteenth-century French comet hunter Charles Messier. Messier's list consisted of faint, nonstellar objects that could conceivably be confused in appearance with a distant comet.

For many years the glow was considered to be a cloud of gas because of its appearance in a telescope. It was then known as the Andromeda nebula. The word *nebula* is derived from a Latin word meaning "cloud," but this proved to be an incorrect term.

A 1912 analysis of the light from the glow revealed the glow to be the light of unresolved stars, stars so far away they could not be seen as individual stars. The Andromeda nebula was a gigantic star cluster. This solved one problem, but created another.

Was M31, the Andromeda glow, a distant cluster of thousands of stars within our Milky Way galaxy, or was it a cluster of billions of stars far outside our Milky Way galaxy? It would look the same to us in either case.

If the latter were true, then we would have to ask if M31 and other similar telescopic objects not visible to the naked eye represent star systems on the scale of the Milky Way itself? Was it possible the Milky Way was not the entire universe, but only one galaxy in a universe of galaxies? The answer could only be found in a measurement of the distance to M31.

In 1923, the American astronomer Edwin Hubble (1889–1953) used the one-hundred-inch Mount Wilson Observatory reflecting telescope, and a cepheid-type variable star, to measure the distance of M31. Hubble found a cepheid variable in M31 and measured its period. Knowing its period, he knew how bright the star would appear at a set distance. He was then able to determine how far away the cepheid star in M31 would have to be to appear as faint as it did. At a calculated distance of 800,000 light years, the cepheid and M31 were too far away to be a cluster of stars belonging to the Milky Way. M31 was another Milky Way, another galaxy. It became the Andromeda galaxy.

In 1952, a mistake was discovered in one of the assumptions used in calculating the distance to the Andromeda galaxy. The cepheid variable stars used to measure the distance to the Andromeda galaxy were discovered to be much brighter than originally believed.

New calculations moved the Andromeda galaxy to a distance of 2.2 million light years and more than doubled its size. The light we see tonight from the Andromeda galaxy left at a time when a prehomo sapien was the most advanced earthly primate. Even at this incredible distance in space and time, the Andromeda galaxy remains the closest of the large neighboring galaxies. In less than half a century the Milky Way was reduced from its position as the whole universe to just one galaxy among billions. In less than the same half century the visible universe was reduced from its position as the whole universe to just one view among many.

It was also discovered that the visual universe was not the only universe. In the year 1800, we received our first clue that "seeing" was more "than meets the eye." William Herschel discovered the existence of infrared energy waves from the sun to which the optical nerve of the eye was insensitive. Ultraviolet energy waves, equally undetectable by the optical nerve, were discovered the following year. There was a universe defined by wavelengths longer than and shorter than the wavelengths of light.

The most revealing "other universe" discovery was made by Karl Guthe Jansky (1905–1950), a communications engineer for Bell Labs. Jansky's problem was to identify the cause of static affecting the company's newly developed transatlantic shortwave radio-telephone system. Jansky discovered that the source of the static was the Milky Way galaxy. The strongest static was coming from the center of the galaxy in the direction of the constellation Sagittarius.

Prior to this discovery, it was not known that very long wavelength radiation was able to penetrate the earth's atmosphere. Most radiation from the universe, except for light, was thought to be absorbed or reflected back into space. The human body never evolved an organ sensitive to radiation in the longer wavelength range of centimeters to meters, so we were in a sense left in the dark about this part of the universe.

"Radio astronomy," an entirely new branch of astronomy, developed as a consequence of Jansky's discovery. Its name is derived from the equipment, techniques, and language borrowed from the radio communication industry that existed at the time of its early development. Radio astronomy added significantly to our knowledge of the universe because it allowed astronomers to detect and study objects that would otherwise have remained invisible to us in ordinary light. The vast clouds of cool hydrogen gas that permeate our galaxy are too cool to radiate light, but they can be detected by radio telescopes. The detection of these clouds has led to a fuller understanding of the structure and dynamics of the galaxy.

One of the oldest problems to be solved by astronomers is that of the origin of the universe. It continues to be an ongoing process that probably started with the first humanoid to lie on its back under a crystal clear, dark night sky. The most accepted answer today to the question of the origin of the universe is the Big Bang theory. This was a phrase given by astronomer George Gamow (1904–1968) and associates in 1948 to their scenario for the creation of the universe. It was simply their version of how the accumulated observations should be assembled.

The Big Bang theory was historically initiated by mathematical implications arising from Einstein's 1916 general theory of relativity. It was given strong support in the 1920s by Edwin Hubble's empirical evidence of galactic redshifts. The redshift is a Doppler effect in light that is analogous to the familiar Doppler effect in sound. When a sound producing object is approaching a listener, it seems to emit a higher pitch sound than when it is moving away from the listener, or when it is stationary with respect to the listener.

A study of the Doppler effect in the light of distant galaxies showed they were all moving away from the Milky Way. And the farther away the galaxies were, the faster they were moving away. These observations were consistent with the observations expected from an expanding universe. But the redshifts of distant galaxies, surprisingly, do not put the Milky Way in a central position in the universe. All observers, in all galaxies, at all locations in an expanding universe, would observe the same Doppler effects. Our expanding three-dimensional, spacetime universe is analogous to the two-dimensional universe of the surface of an expanding balloon. The observations in both situations are the same.

Hubble's redshift measurements were the first observational clues that the universe was expanding. They are remnants of the cosmic fireball still observable today as the universe continues to expand following the explosion at the beginning of time and space. The temperature of the universe is also a remnant of the Big Bang. As the universe expanded, it also cooled. A critical test of the Big Bang theory was found in a calculation for the predicted temperature of the present universe.

Bell Labs radio astronomers Arno Penzias and Robert Wilson accidently discovered the predicted 2.7-degree Kelvin primordial background radiation left over from the Big Bang while trying to eliminate an unwanted source of noise from a microwave receiver.

Their measurement in 1965 of a uniform temperature throughout the universe equal to the predicted value was the strategic observational clue that the universe had experienced a violent beginning. It was the deciding factor in favor of the Big Bang. The Big Bang was the theory that best fit the facts. The importance of the discovery won Penzias and Wilson a Nobel Prize in 1978.

According to the Big Bang theory, the universe began fifteen to twenty billion years ago with an explosion—a bang. It was the greatest explosion possible since all of time and space were created out of the explosion. It was the ultimate explosion—the Big Bang.

Radiation dominated the early seconds of the universe, but within two minutes simple hydrogen and helium nuclei began to form. The rapidly expanding matter of the universe eventually fragmented into enormous clouds of gas, which further fragmented into stars. These galaxies of stars began to form when the universe was about one billion years old.

Massive stars within the galaxies evolved, created more complex nuclei, and shared these nuclei with surrounding space through "little bang" supernova explosions. The rags of the old stars became the riches of the new stars, and any planetary system accompanying them. A fragmentation ten billion years ago produced the Milky Way. A fragmentation occurring five billion years ago within the Milky Way produced the sun, our solar system, and eventually living, intelligent human beings attempting to decipher the events that produced them.

Is modern astronomy the result of only Western astronomers? The contribution of India's Megh Nad Saha's ionization theory to an understanding of the spectra of stars, and ultimately the chemistry and physics of stellar atmospheres, would seem to say no. The contribution of Subrahmanyan Chandrasekhar's research into the evolutionary paths of old stars would seem to reinforce that no. But a closer inspection of their achievements in astrophysics would seem to say otherwise. Saha and Chandrasekhar were Eastern Hemisphere astronomers working in Western research environments.

In the 1920s the astrophysicist Megh Nad Saha from the University of Calcutta in British-dominated India applied the theory of the make-up and behavior of atoms to the spectra of stars. Saha was able to successfully explain the different appearance of the fingerprint lines of the same elements in different stars as the result of the different atmospheric temperatures of the stars. But Saha's theory of stellar spectra, to which his name is inextricably linked, was exported for publication. The Saha ionization theory was first announced in a British scientific journal.

Another major astrophysical theory developed by an Indian was done entirely outside of India. Subrahmanyan Chandrasekhar, the most productive twentieth-century Indian astronomer, did all of his astrophysical research at the Yerkes Observatory of the University of Chicago. His research led to an understanding of the death of old stars. The derivation of the "Chandrasekhar limit" established 1.4 solar masses as a dividing line in stellar evolution. Stars like the sun with masses below the 1.4 solar mass limit end quietly by collapsing to form white dwarf stars. Stars above the limit end in cataclysmic supernova explosions that leave behind neutron stars and black holes. Chandrasekhar received the most prestigious and coveted form of recognition for scientific achievement. He shared the 1983 Nobel Prize in physics.

The Eastern and Western Hemispheres of the earth are culturally different in many ways. Some of the differences are subtle. A Westerner must often interact with an Eastern culture to recognize them. Some of the differences are not so subtle. They are seen by every tourist from the West. The differences are quite clear in terms of the history of the astronomical interests of the two hemispheres.

Lockyer's third stage involving the pursuit of astronomy for the acquisition of knowledge without regard to its applications dominates the Western Hemisphere's interest in the night sky. The history of research astronomy is dominated entirely by developments in the West. The contributions from the East have been meager. Lockyer's second stage involving the utilization of celestial phenomena to serve the practical needs of culture has dominated the Eastern Hemisphere's interest in the night sky over a long period of time. Astrology is a force that has dominated the thinking and activities of the East. Its influence in the West has been considerably less.

What kept the Eastern Hemisphere from a dedicated interest in the pursuit of pure astronomy? Why did the Eastern Hemisphere not make significant progress beyond astrology? Why did pure astronomy flourish only in the Western Hemisphere? Here is a suggestion. Astrology satisfied the religious and philosophical thinking of the cultures of the Eastern Hemisphere, while astronomy satisfied the religious and philosophical thinking of the cultures of the Western Hemisphere.

The East's traditional religious and philosophical schools of thought can be broadly characterized as cyclical. The Shiva creation myth provides a representative example of the time-honored depth of this attribute.

In the Eastern Hemisphere, the universe was seen by the ancient Hindus as the one-night cosmic dream of Brahma. Brahma's dream of 4,320 million earth years began with the dance of the four-armed god Shiva, the Lord of the Cosmic Dance and God of Creation and Destruction. Shiva's dance began each new cosmic dream and each new universe. In the Hindu tradition, Brahma's waking will eventually end the dream and the universe. But after Brahma's dream, another sleeping and waking god will appear. And Shiva will dance again to create another universe. The ancient Hindu creation myth presents a universe of cosmic cycles.

There is another culturally influencing element operating in the East. Traditional Eastern religious and philosophical schools of thought can also be characterized as animistic in their world-view. They are "why" questioners seeking a designing intelligence beyond a facade of sense perception. They approach an explanation of the dualistic universe they perceive in terms of underlying intentions and purposes.

Traditional Eastern religious and philosophical thought seeks to spiritually escape a confining dependence on the body and the physical world. There is, understandably, little emphasis or interest in seeking a comprehension of that from which one is attempting to escape. There was no need or desire in the Eastern Hemisphere to ask scientific questions of the universe. The East was satisfied with astrology because astrology provided a satisfying scheme of the universe. Astrology provided a mirror image of an animistic universe.

Astrology deals with cycles and portrays the believer as a necessary interactant in those cycles. The believer in astrology is made an integral component of a cyclical, purposeful, animistic universe just as with Hinduism, Buddhism, and Taoism, the major religions of the Eastern Hemisphere.

Hinduism, with its emphasis on karma and dharma, is more attuned to astrology than astronomy. Karma, the belief that actions in the present affect future existences as the eternal soul goes through a cycle of births, deaths, and reincarnations, is only an extension of the recurring cycles of the celestial domain. And dharma, the fulfilling of one's duty in life as dictated by social background and custom, is only an extension of the fulfilling of one's destiny caused by the influence of the appearance of the celestial domain at birth.

The individuality of the believer in astrology and Eastern religions and philosophies is suppressed and made a part of the universe. The believer is a component of universal cycles.

Traditional Western religious and philosophical schools of thought can be broadly characterized as linear. The ancient Greek creation myth provides a representative example of the time-honored depth of this attribute.

In the Western Hemisphere, the universe was seen by the ancient Greeks to have evolved from a disordered state of unformed matter. In the beginning there was chaos. This confused universe formed into an orderly arrangement of stars and planets. A cosmos was created. The gods of Olympia were created out of that ordered arrangement of matter. In Greek mythology, as opposed to Hindu mythology, the gods were not responsible for the universe, nor were they superior to the universe. They were created by the universe. And the universe had only one beginning. It was a linear universe.

Like the East, there is another culturally influencing element in operation in the West. Traditional Western religious and philosophical schools of thought can also be characterized as mechanistic in their world-view. Westerners are "how" questioners seeking precise mathematical relationships between sense perceptions. They approach an explanation of the universe in terms of physical cause and effect relationships.

The major religions of the Western Hemisphere are not cyclical like those of the East; they are linear in thought and purpose. Judaism, Christianity, and Islam start with an event and proceed toward a promised ending event. In Western religions the individuality of the believer is paramount. The universe is a stage on which to achieve individual salvation; the believer has only one curtain call.

The pursuit of pure astronomy, like traditional Western religions and philosophies, is a linear activity. The Western astronomer studies the universe as an outsider and proceeds toward the end of acquiring a knowledge of this universe in much the same way as a religious believer studies sacred literature and applies the logic of cause and effect to acquire a knowledge of religious truth. The astronomical model of the universe that has evolved in the West is also linear. It began some fifteen billion years ago with the Big Bang and is proceeding toward a cold motionless end.

The dark sky legacy of the Eastern Hemisphere is different today from the dark sky legacy of the Western Hemisphere. It is understandable that astrology dominates the Eastern Hemisphere, with its traditional religious and philosophical emphasis on cycles. It is equally understandable that astronomy dominates the Western Hemisphere, with its traditional religious and philosophical emphasis on linearity.

The present situation will not continue into the future. The cultures of the Eastern and Western Hemispheres will merge and develop into subcultures. There will still be astrologers. There will still be astronomers. The impetus for this fusion will be found in the pursuit of space and the economics of East-West trade. The Western Hemisphere will eventually westernize the Eastern Hemisphere. The process has already begun. With this, another change will come. The dark sky legacy of the Western Hemisphere will become the dark sky legacy of the earth.

Selected Bibliography

Dark Sky Legacy was not in my mind twelve years ago when I began to assemble the material that has resulted in this book. I therefore confess to not having kept a careful record of many of the sources for the information presented. The following bibliography is highly selective, and regrettably incomplete. However, it will serve the purpose of leading an adventuring reader in the right direction.

Books

Asimov, Isaac, *Words From the Myths,* Boston: Houghton Mifflin Company, 1961.

Bradford, Ernle, *Christopher Columbus,* London: Michael Joseph Ltd., 1973.

Calder, Nigel, *The Comet is Coming: The Feverish Legacy of Mr. Halley,* New York: Penguin Books, 1982.

Christianson, Gale E., *This Wild Abyss: The Story of the Men Who Made Modern Astronomy,* New York: Macmillan Publishing, 1978.

Cornell, James, *The First Stargazers: An Introduction to the Origins of Astronomy,* New York: Charles Scribner's Sons, 1981.

Ferris, Timothy, *The Red Limit: The Search for the Edge of the Universe,* New York: Bantam Books, 1979.

Gauquelin, Michel, *The Scientific Basis of Astrology: Myth or Reality,* New York: Stein and Day, 1969.

Gauquelin, Michel, *Dreams and Illusions of Astrology*, Buffalo, N.Y.: Prometheus Books, 1979.

Goldstein, Thomas, *Dawn of Modern Science*, Boston: Houghton Mifflin Company, 1980.

Gropman, Donald, *Comet Fever: A Popular Review of Halley's Comet*, New York: Simon and Schuster, Inc., 1985.

Hadingham, Evan, *Early Man and the Cosmos*, London: William Heinemann, Ltd., 1983.

Hawkins, Gerald S., *Mindsteps to the Cosmos*, New York: Harper and Row, 1983.

Hawkins, Gerald S., and John B. White, *Stonehenge Decoded*, New York: Harper and Row, 1965.

Hoyle, Fred, and N. C. Wickramasinghe, *Diseases from Space*, New York: Harper and Row, 1980.

Koestler, Arthur, *The Sleepwalkers*, New York: Gosset and Dunlap, 1963.

Krupp, E. C., editor, *In Search of Ancient Astronomies*, New York: McGraw-Hill, 1978.

Krupp, E. C., *Echoes of the Ancient Skies: The Astronomy of Lost Civilizations*, New York: Harper and Row, 1983.

Kuhn, Thomas, *The Copernican Revolution*, New York: Vantage Books, 1957.

Lockyer, J. N., *The Dawn of Astronomy: A Study of the Temple Worship and Mythology of the Ancient Egyptians*, London: Cassell, 1894.

Manuel, Frank E., *A Portrait of Isaac Newton*, Washington, D. C.: New Republic Books, 1979.

Meadows, A. J., *Science and Controversy: A Biography of Sir Norman Lockyer*, Cambridge, Mass.: The MIT Press, 1972.

Oberg, James E., and R. Alcestis, *Pioneering Space: Living on the Next Frontier*, New York: McGraw-Hill, 1987.

Parker, Derek and Julia, *The Compleat Astrologer*, New York: McGraw-Hill, 1971.

Pioneering the Space Frontier: The Report of the National

Commission on Space, New York: Bantam Books, 1986.

Powers, Robert M., *Mars: Our Future on the Red Planet,* Boston: Houghton Mifflin, 1986.

Serviss, Garrett, *Astronomy with the Naked Eye,* New York: Harper and Brothers, 1908.

———, *Round the Year with the Stars,* New York: Harper and Brothers, 1910.

Weinberg, Steven, *The First Three Minutes,* New York: Bantam Books, 1979.

Withey, Lynne, *Voyages of Discovery: Captain Cook and the Exploration of the Pacific,* New York: William Morrow and Company, Inc., 1987.

Magazines

Astronomy, published by AstroMedia, 1027 North 7th Street, Milwaukee, WI 53233.

Griffith Observer, published by Griffith Observatory, 2800 East Observatory Road, Los Angeles, CA 90027.

Sky and Telescope, published by Sky Publishing Corporation, 49 Bay State Road, Cambridge, MA 02138.